INTO THE GREAT EMPTINESS

Also by David Roberts

INTO THE
GREAT
EMPTINESS

PERIL AND SURVIVAL ON THE GREENLAND ICE CAP

DAVID ROBERTS

W. W. NORTON & COMPANY
Independent Publishers Since 1923

Publisher's Note

W. W. Norton & Company, Inc., wishes to thank the Estate of Sir Martin Lindsay* for its kind permission to quote substantive passages from *Those Greenland Days*, Lindsay's own account of the British Arctic Air Route Expedition to the east coast of Greenland in 1930–31, which is the thrilling subject of *Into the Great Emptiness* by David Roberts. Gino Watkins, the head of the expedition, was acknowledged by his colleagues as a brilliant explorer and leader of men, though he was the youngest member of the group. But Watkins died in a kayaking accident shortly after BAARE left Greenland, leaving no account of what he and his men had accomplished, not even notes. Hence the importance of books and recollections by other BAARE members. David Roberts relied extensively on the account in *Those Greenland Days* in his research, because of the way in which Sir Martin recounted his careful observation of the very many telling details that allow the reader to understand what it was really like to be on such an expedition in Greenland.

* Lieutenant-Colonel Sir Martin Lindsay of Dowhill, Bt., CBE, DSO, (22 August 1905–5 May 1981) was a British Army officer, polar explorer, politician, and author.

For information about permission to reproduce selections from this book, write to
Permissions, W. W. Norton & Company, Inc., 500 Fifth Avenue, New York, NY 10110

For information about special discounts for bulk purchases, please contact
W. W. Norton Special Sales at specialsales@wwnorton.com or 800-233-4830

Manufacturing by Lakeside Book Company
Book design by Patrice Sheridan
Production manager: Julia Druskin

Library of Congress Control Number: 2022008994

ISBN 978-1-324-08637-6 pbk.

W. W. Norton & Company, Inc., 500 Fifth Avenue, New York, N.Y. 10110
www.wwnorton.com

W. W. Norton & Company Ltd., 15 Carlisle Street, London W1D 3BS

10 9 8 7 6 5 4 3 2 1

For Stuart Krichevsky—

Wisest of agents,

Most loyal of friends,

And for me, a reason to keep going

CONTENTS

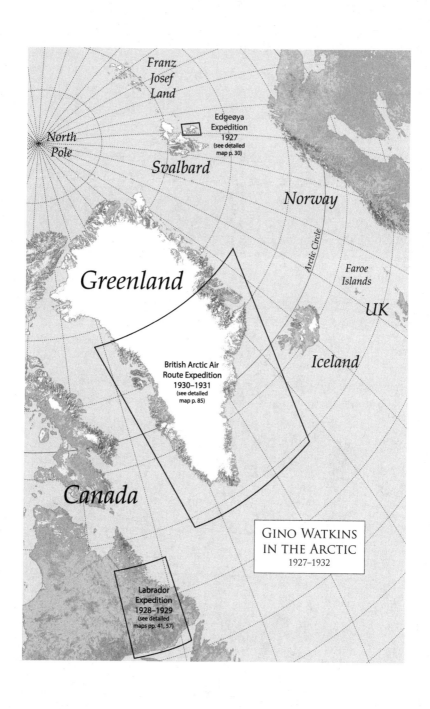

Franz
Josef
Land

Edgeøya
Expedition
1927
(see detailed
map p. 30)

North
Pole

Svalbard

Norway

Arctic Circle

Greenland

Faroe
Islands

UK

British Arctic Air
Route Expedition
1930–1931
(see detailed
map p. 85)

Iceland

Canada

GINO WATKINS
IN THE ARCTIC
1927–1932

Labrador
Expedition
1928–1929
(see detailed
maps pp. 41, 57)

INTO THE GREAT EMPTINESS

PROLOGUE

The Man on the Ice Cap

JAMIE SCOTT AND Quintin Riley outfitted two sledges, rounded up the huskies, and harnessed them into their leads. The mission on which they were about to set out was in one sense a familiar one. Members of the team had performed it three times the previous autumn. But this time a hectic urgency hung over their preparations. A single man, having passed the winter alone, lay waiting in the big domed tent that served as the expedition's Ice Cap Station. August Courtauld, who had volunteered for the solo vigil at the beginning of December, needed to be relieved from a duty the likes of which had never before been attempted anywhere on earth.

The date was March 1, 1931. From the expedition base camp in a nestling fjord on the east coast of Greenland, Scott and Riley needed to climb through the steep glacial headwall that had given the team fits on every previous foray, then head across the blank immensity of the ice cap, following the red flags planted in the snow at every half mile, until they reached the station, 130 miles inland and 8,200 feet above sea level. The five men who had pioneered the route in August 1930 had spent two weeks in the effort. Scott and Riley hoped their own relief journey would take no longer.

The expedition had chosen as its clunky title the British Arctic Air Route Expedition, acronymed as BAARE. It comprised fourteen men, all but one in their twenties, most of them students or graduates from Cambridge University. Among them, only Courtauld had been to Greenland before, on a pair of modest probes of the ice pack–tormented east coast. But Courtauld was not the expedition leader. That man was Henry George Watkins, nicknamed "Gino" since childhood (despite no family ties to Italy). The BAARE was entirely Watkins's campaign, a vastly ambitious exploratory attack on the largest island in the world— and one of the least known parts of the globe. Gino devised seven separate mini-expeditions to be carried out across the span of sixteen months. By 1930 he was already the veteran of two previous probes of the geographical unknown that he had organized himself, to Svalbard in the high Arctic and Labrador on the eastern coast of North America. Yet at the outset of the BAARE, Watkins was only twenty-three years old—the youngest of all its members. His youth notwithstanding, Gino's assault on Greenland would constitute the most daring and fruitful British expedition to the Far North during the previous half century, comparable in its stature among the nation's polar exploits only to the Antarctic missions of Robert Falcon Scott and Ernest Shackleton two decades earlier.

More than a century after their deaths, Scott and Shackleton are enshrined as legendary explorers. Yet, as even his admirers admit, Gino Watkins has lapsed into the limbo of the "forgotten hero." The reasons are several. By dying in the early days of his fourth expedition in a fluke accident at the age of twenty-five, he blazed a track across the heavens that winked out like that of a meteor that never reaches earth. Unlike Scott and Shackleton, Watkins was too busy planning each subsequent jaunt to write more than a couple of dry articles for the *Geographical Journal*. The expedition books were written by his teammates.

Gino remains in some sense the Mozart of Arctic endeavor, the child prodigy who died before his full genius could flower. His schemes were wildly, even arrogantly, bold. Yet he backed up his boasts with an adaptive skill that far surpassed Scott's or Shackleton's (who, despite their years in the ice and snow, never learned how to build an igloo or how to hitch dogs to their sledges). With three major expeditions under his belt by the age of twenty-four, Watkins (unlike Scott and Shackleton) had never lost a man.

As multipronged as the BAARE's goals were, its core feature was the Ice Cap Station. The 660,000-square-mile sprawl of permanent ice that covers all of Greenland but for its thin coastal margins makes it, with its partner in Antarctica, one of the two greatest ice sheets in the world. And its mean altitude of 7,000 feet renders it one of the coldest and most forbidding regions on earth. By 1930, the mystery of the atmosphere above the ice cap was seen by scientists as promising the key to the strange ocean currents that swept the Atlantic, and indeed to the climate of the eastern Arctic itself. The Inuit, who had colonized the coasts of Greenland centuries, maybe millennia, before Norsemen discovered the island, never (as far as we know) dared venture onto the ice cap. Inuit legends populated that interior with monsters, giants, and malign spirits, who despite their remote domain interfered at all turns with human affairs. These were not vague, ethereal beings, but vividly physical creatures, such as the *timertsit*, a giant with a massive lower jaw from which hung a stone lamp and heating bowl for cooking humans alive.

The ice cap was first traversed in 1888 by a team led by the great Norwegian explorer Fridtjof Nansen. By 1930, it had been crossed several more times by Europeans on even longer routes. But no one had ever tried to winter over in that forbidding center, let alone manage a weather station through twelve continuous months.

Besides that meteorological program, Gino had another scheme in view, the one that supplied the "air route" in the BAARE title. He had noted (as had others) that the shortest line from Britain or western Europe to North America was an arc across the Arctic. Having already earned his own pilot's license, Watkins now laid out a plan to pioneer such a route. In 1930, the idea of regular passenger flights across the vast expanse of the Atlantic Ocean (much less the Pacific) seemed hopelessly futuristic. Charles Lindbergh had completed his daring solo flight from New York to Paris only three years earlier, carrying the bare minimum of food and survival gear to offset the necessary fuel.

Any air route from the Old World to the New, then, would have to proceed in short jaunts hopscotching between landing fields not more than 500 miles apart. So Gino sketched a seven-part itinerary: London to the Faroe Islands to Reykjavík to Angmagssalik on the east coast of Greenland; then across Greenland to Disko Bay on the west coast, from which another leg would cross Baffin Island to land near Cape Wolstenholme on Hudson Bay; and thence to Fort Churchill and Winnipeg. The great unknown factor for future pilots along this route lay in the high Greenland ice cap. Thus the year-round weather station might also monitor the storms and winds and snowfall that could threaten or even doom an Arctic air route. Watkins had bought two de Havilland DH60G Gypsy Moth airplanes, carried them in crates to Greenland, and assembled them at base camp. In his grand plan, toward the end of expedition, he would pilot the maiden flight himself from Angmagssalik to Winnipeg.

Although scientific and commercial motives seemed to dictate the aims of the BAARE, closer to Gino's heart (and to those of his teammates, of whom only one was a scientist) was the old itch to explore, to discover worlds that no one before had ever seen. That itch had first been awakened at nineteen, when, still a freshman, he had attended a lecture by Cambridge don Raymond Priestley, who told stories of his days with both Scott and Shackleton in the Antarctic. Leaving the lecture with his school chum Quintin Riley, Watkins blurted out, "I think we'd better go to the Arctic."

Four years later, the chance to venture where human beings had never been, to tear the veil from the unknown, was all that Gino Watkins lived for.

* * *

On March 1, Scott and Riley could not have gotten off to a worse start. The passage through the glacial headwall, only two miles from base, nearly stopped them cold. The previous August, teams had seen the dogs' feet cut and torn on the jagged ice there, and it turned out that the supply of canvas canine booties had been understocked by 75 percent. Even with men hauling along with dogs, the sledges had slid back down the slope or capsized. It finally took a winch and the installation of fixed ropes to get the craft and loads up the headwall, which in disgust the team nicknamed Buggery Bank (euphemized in print as Bugbear Bank).

Now Scott and Riley had to reckon with the weeks in February of gales that had worn the whole surface (even beyond Buggery) "as hard as marble" with "even more formidable ditches and banks." Having managed to crest the headwall, the two men pushed on to the Big Flag depot, fifteen miles from the coast. This massive supply cache, crowned with a tall flag, was the key to the whole long route to the Ice Cap Station. But only a mile beyond Big Flag, Scott saw his sledge, battered by rock-hard ice, break in half. There was nothing to do but retreat to base camp, which the two disheartened men reached the next day.

Scott asked Watkins to assign another man for the second attempt. With three sledges instead of two, the trio could afford to lose one and still forge on. Gino chose Martin Lindsay, the tallest man in the team and one of the strongest, for the task.

On March 4, the three men set out again. This time they did not even reach the Big Flag depot. The morning had been beautiful, but by midday they were engulfed in a snowstorm so dense they feared they would lose the track. Only three miles beyond Buggery Bank, they called a halt and pitched their lightweight tent. Certain they would reach Big Flag and the camping gear deposited there that day, the men had brought only sleeping bags and a light lunch of biscuits and chocolate. At the last minute, on a whim, Riley had added thirty more biscuits to his lunch bag.

The storm raged on for four days. With no boxes or gear to weigh down the skirts of the tent, the men feared that their shelter would be ripped away if they even dared to try to pack up. They had no Primus stove, so they melted snow cup by cup over a candle flame. As Lindsay reported, "This was a tedious business and it was also a dirty one, for our hands and faces were soon black with soot. The biscuits got fewer and fewer."

Big Flag was only two miles away—"on a fine day we could have seen [it] from where we lay"—but on the fourth morning the trio gave up. The sledges were utterly buried under new snow. With their only tools, a knife and a cooking pot, they dug out a single sledge and harnessed all the dogs to it. They collapsed the tent and left it there. Sinking knee-deep even on snowshoes, with the dogs so helpless they "might almost have been swimming," Scott's party at last abandoned the sledge and made a desperate dash back to base camp, arriving late at night.

A week had been wasted getting nowhere. It was not until March 9 that the relief team launched their third attempt.

Meanwhile, 130 miles away, Augustine Courtauld (known to all his friends as "August") was beginning to wonder if he had been abandoned. It had not been part of the original plan to have him monitor the weather station solo through the winter. Watkins's scheme called for rotating teams of two men each to man the station for at most two months at a time. That design worked for the first two pairs, from August through November 1930, though the occupants found their isolation on the featureless plateau spooky and disturbing.

The third resupply, a six-man team led by F. Spencer (Freddy) Chapman, overcame one obstacle after another, as fierce storms and whiteout marooned them in their tents for days at a time, while the dogs grew weak from the toil and barely adequate rations. Having set out on October 25, the team made such poor progress that on November 11 Chapman sent three men back to base to save fuel and food. The remaining trio arrived at the Ice Cap Station only on December 3. A trip that had been expected to take two weeks ended up stretching across five, as darkness and cold intensified day by day. On December 3, knowing they were near the station but crisscrossing the featureless waste unable to find it, the party grew frantic from "frightful disappointment." At last, navigating in the dark by Arcturus and Venus, Chapman stumbled upon the domed tent and recording gauges "a hundred yards to the left of where they should have been." The occupants, "at first refusing to believe it could be the relief party arriving," were overjoyed to be met by their comrades.

By December 3, the relief team had only one day's rations left for the dogs. The dash back to base camp would be an ordeal under the best conditions. The plan had been for August Courtauld and Lawrence Wager—the team's geologist and its best mountaineer—to occupy the station through the next shift. But the relief party had had to break into food boxes meant for the third pair of monitors. There were not enough provisions and fuel left for another shift, especially given the difficulty faced by the third resupply mission in just getting there.

It looked as though the station would have to be abandoned. The central campaign of the BAARE teetered on the verge of wreckage.

At this point, Courtauld made an astounding offer. He would be willing to serve the next shift alone. For one man, food and fuel ought to be adequate through mid-March, or, stretching the supplies paper-thin, even into early May.

The pair who had just been relieved at the Ice Cap Station were aghast. Through the night of December 3–4, ensconced in the domed tent, all six men argued back and forth. The station pair rehashed a couple of incidents during their stint in which one man had gotten into trouble, only to be helped to safety by his partner. The psychological toll alone had been extreme. It seemed to them an unimaginable trial for a solo occupant.

In his mild-mannered, almost diffident way, Courtauld persisted in his offer. He rather liked being alone, he said. He cited Gino himself on trappers in Labrador who wintered alone in remote camps without mishap. Courtauld was utterly committed to the expedition's goal: in Chapman's telling, he argued that "the winter, when nobody had ever lived on the ice cap, was the most important season for recording the weather."

Yet another consideration mitigated against the brazen proposal. On the November journey up to the station, Courtauld had developed frostbite of both big toes and fingers, the latter swollen so badly that he couldn't undo his buttons. Almost perversely, he now used those infirmities to further his case: he didn't relish the thought of getting worse frostbite on the trek back to base.

At last the other five men gave in. On December 5, Courtauld saw his teammates sledge away. "It was bitterly cold and I didn't watch them long," he wrote in his diary. "Coming out an hour later I could just see them as a speck in the distance. Now I am quite alone. Not a dog or even a mosquito to look at."

Three months later, on March 9, 1931, Scott, Riley, and Lindsay launched their third effort to get to the station and relieve Courtauld. They packed their anxiety into their sledge-loads. Lindsay and Riley knew what it was like to man the Ice Cap Station, for they had served as the first pair of monitors for five weeks after August 30, 1930. They had found the accommodations "very snug and comfortable"; yet as their stint drew toward its end, the inevitable malaise seized their spirits, and they spent many an hour discussing whether they could hike back to base camp without dogs or sledges if no relief party came.

Yet now, in March 1931, Scott, Lindsay, and Riley could not imagine that their own journey might face greater hardships than those Chapman's team had battled the previous November. But that is what happened.

In November, Chapman and crew had brought along a wireless set, in hopes of setting up radio communications between base and Ice Cap

Station. The device, however, made a monstrous sledge load, the generator alone weighing 150 pounds. When the team reached a supply depot at Flag 56, some thirty miles out, they gave up on the wireless set and left it there. Despite that seemingly secure cache, the set was never found again.

Now Scott's three-man party faced another grim choice. To determine latitude, observations with a theodolite were sufficient—except during storms. But longitude was a far trickier matter. The team had time-set devices, essentially accurate chronometers for determining mean difference from Greenwich. But these gizmos were also bulky, heavy, and fragile; so, mindful of the desperate conditions of the two botched forays, Scott decided to leave them at base camp. The relief party would count on the red flags every half mile, but to know how far west they had gone, they must rely on a time-honored contraption. The sledge wheel was just that: a circular tire lashed to one of the sledges, which counted up the distance traveled foot by uncertain foot. Crude though it sounds, the apparatus had faithfully served many another polar expedition.

By pushing hard on the good days, Scott's team covered half the distance to the Ice Cap Station in eleven days. From there on, though, their advance sputtered. Many of the half-mile flag markers had disappeared, either blown away or covered by drifts. Scott now attempted even more careful theodolite readings, vexed by a half dozen problems, from the nonappearance of the sun to the difficulty of stabilizing the three tripod legs evenly in the snow to thawing the instrument inside the tent to extract its reading. In the lacunae between red flags, the sledges wove right and left of course, adding false distance to the sledge-wheel count.

Scott's plan was now to reach the exact latitude of the station (which had been precisely determined) at a point some miles east of the station, then sweep methodically along that contour to a point well beyond where the domed tent ought to be, recheck the latitude, then sweep back, the three men taking separate routes hundreds of feet apart. On

March 27 the trio camped at a point they judged to be nine or ten miles east of the station, smack on the right latitude. The telltale marker for the station was a big Union Jack hoisted on a sturdy pole high above the tent. In all the white smear of the world, that landmark ought to stand out plain, visible from a distance.

But now the men swept back and forth for almost *three weeks*, though confined to their tent by storms on half of the days, and found nothing. Hope gave way to desperation, then to despair. It even occurred to the searchers to blame Courtauld for his disappearance. Wrote Lindsay, "For the last two miles we were all the time expecting to see the station over the next rise. We were confident that we had not passed it by more than half a mile; but we had seen no sign of it. It was difficult to understand how Courtauld could have come to let the station get so drifted over—unless he had met with some misadventure."

At last they gave up. On their demoralized run back to base, they had to kill two of the huskies to feed the others. Another dog gave birth to a puppy, promptly devoured by other dogs. The men's own rations had long since been cut to two-thirds.

It was not until April 17 that the team got back to base, "having wandered for forty days in the wilderness," as Lindsay put it, echoing the Gospel of Matthew. Entering the hut, they broke the bad news to their stunned teammates.

Alone among the men gathered at base, Watkins remained sanguine. He laid out no blame for the team's failure, but at once started to put together a fourth team, to be led by himself. "He was convinced that Courtauld was all right," recorded Lindsay, "and nothing could shake his faith."

Through December 1930 and January 1931, Courtauld had adjusted well to his solitude. His only chores were to crawl out of the tent at regular intervals six times a day and read and record the gauges, and while that gave structure to his life, during storms those errands could become brutal tasks. Yet lying in his sleeping bag, he felt almost lazy, as he contemplated the rigorous work the rest of the BAARE team must be performing.

Courtauld voraciously tackled the library of books stashed in the big tent, reading and rereading his favorites, such as *The Forsyte Saga* or *Jane Eyre*. In his thoughts and his diary, he sent loving messages to Mollie Montgomerie, the sweetheart who had become his fiancée shortly before the expedition. Yet as the winter crept on and darkness reigned over the plateau, he grew anxious about the prospect of relief. Unwilling to dwell in ambiguity, he decided that March 15 would be the deadline by which he would expect the relief team to arrive, and at first he adjusted his food intake and the supply of precious tobacco for his pipe accordingly (with a small extra allowance in case his deadline proved wrong).

Gino himself had designed the station. The big domed tent in which the monitors lived was nine feet wide by seven feet tall, its walls supported by strong bamboo wands. Over that a separate fly was draped, guyed tight so that an ample space between tent and covering ensured against snow building up on the walls. A single metal pipe two inches in diameter extended from the roof, delivering life-saving air if the tent somehow got smothered in snow. Around the tent, all kinds of measuring devices—a comb to gauge cloud speed, a cup anemometer to record the force of wind, snow-depth gauges, a maximum-minimum thermometer, and a barograph—occupied their own stations in little snow-block shelters a few feet away from the living quarters.

The crucial design innovation Watkins had made was to provide no door in the walls of the tent, but instead a trap door in the floor that gave onto a subsurface snow tunnel along which the monitor of the day would crawl and emerge from its mouth near the gauges. This worked superbly for the first two teams and through Courtauld's first few weeks, for in the worst storms none of the blown snow made its way inside the tent. But that same innovation would prove to be the station's lethal flaw.

In January a hurricane piled snow so thick in the far end of the tunnel that Courtauld had to try to dig it out from inside. When he couldn't manage that, he dug a sideways tunnel and emerged next to the gauges. Then he stuffed an empty provision box into place in the new tunnel exit, hoping to forestall another blockage. Despite the fix, that glitch in Watkins's station design plunged Courtauld into dark broodings. In his diary, for the first time he contemplated the possibility of dying during his vigil.

March 15 came and went with no sign of a relief party. After March 27, during the weeks that Scott's team swept the terrain where the station, they were sure, must stand, Courtauld never heard a sound. Before then, since late February, each time he had emerged to read the gauges, his eyes had swept the eastern horizon hoping to spot the distant speck of the arriving team. In the many hours of down time inside the tent, he faithfully recorded his doings—as well as his hopes and fears—in his diary. Like all Englishmen of his generation, Courtauld had grown up with the legend of the deaths of Robert Falcon Scott's five-man party returning from the South Pole in February 1912. Only the discovery of Scott's diary nine months later had revealed the story of the team's demise in all its tragic glory.

After the January blockage of the original tunnel, each time Courtauld crawled out the side tunnel, it too filled with blowing snow. He pondered carving a shaft straight up from mid-tunnel to the surface, but, realizing that some eight feet of snow covered it, judged the task impossible. In the side tunnel he was soon reduced to a tight caver's crawl, and from that prone position it took all his strength to push open the hatch door he had improvised with the empty ration box.

On March 17, while Scott's trio were still struggling up the plateau near the halfway point of the long route, disaster struck. Another fierce gale swept the polar plateau, piling new drifts everywhere. One of them covered the ration-box escape hatch. When Courtauld tried to push it out, he couldn't move it an inch. The weight of snow on top of it had turned the tunnel into a locked vault. During the next few days he tried to whittle away at the blockage with a knife, but the plug turned to solid ice. "So I am completely buried," he wrote in his diary.

Forget the gauge readings. From now on, all Courtauld could do was lie helpless and wait for the relief party to find him. No wonder that, ten days after his crypt was sealed, Scott's party sweeping the plateau saw no sign of human life. No wonder that Courtauld, entombed in the death trap the Ice Cap Station had become, with snow drifted up to the roof of the tent, heard no sound from the searchers.

The days came and went. He ate meagerly, read, and smoked the pipe Mollie had given him. On April 17, the same day that Scott, Riley, and Lindsay staggered, defeated, back into base camp at midnight, Courtauld finished the last crumbs of tobacco. In his diary he wrote, "There is now precious little left to live for."

ONE

"He Never Discussed
Anything Seriously"

AT THE AGE of eighteen, Gino Watkins nearly killed himself climbing in the Alps. The way he responded to that brush with death reveals the mercurial genius that would spring full-blown in Greenland five years later. And yet. . . .

Gino's father, the fourth successive Henry George Watkins, was a colonel in the Coldstream Guards, a member of the landed gentry who managed slowly to fritter away his fortune, and a keen outdoorsman. One day in 1914 as Gino, aged seven, watched his father and uncle play lawn tennis at the uncle's country house, a footman interrupted the game with a telegram telling the colonel to report to the Allied front in France. Shortly thereafter, Gino sent a letter to the trenches: "My dear Father. How goes it. How many Germans have you killed lately. . . . I should love to join in the fighting. We go fishing in the lake. I hope you are having a very nice time."

The Watkins family lived in a spacious house on Eaton Place in London's upscale Belgravia district, complete with a beloved Nanny Dennis and several servants. (The nanny would stay in the household throughout Gino's lifetime.) Holidays drew the Watkins clan to Dumbleton, the uncle's pastoral estate in Worcestershire ("ponies to ride, a lake with boats"), and to grandmother's Lilliput, on Poole Harbour near the south coast ("picnics on the sand, bathing in the sea, and shrimps with tea"). Yet Gino had been labeled since his birth in January 1907 as, in that classic Edwardian formula, "a delicate child." His mother, née Jennie Monsell, was clearly disappointed in the scrawny infant she had just borne. "Well, Jennie," her own mother offered, looking over her shoulder, "he's got a neat little face."

Because of his delicacy, Gino was held back from boarding school the first year. Yet from infancy on, he displayed a rebellious, even an anarchic, bent. "You couldn't make him eat properly," his mother swore. "He liked to sit at the table in front of the mirror so that he could watch himself making awful faces as he was eating. . . . And once when he was given his bread and butter in bed he squashed it up in his hand and threw it at the ceiling."

After the births of a younger sister and brother, Gino became a tyrant. He forbade his siblings to play with other children, and terrorized them with tales of dead men under the drawing room carpet or condemned prisoners haunting the neighborhood. Gino himself was terrified by ghosts, and couldn't sit through *Peter Pan* without hiding under his seat. Matters were not helped when his father came back from a trip to Russia with a live baby bear in a cage. Gino had been promised a teddy bear, but on first meeting, "Popoff" hurled himself against the cage's netting. The terrified boy lurched backward, colliding with the wall behind him.

Fragile, fearful, or no, Gino was gifted with a wild imagination, and his father's wanderlust seeped into his own spirit. He kept close tabs on the war, and after the Armistice, when Watkins *père* stayed on in France, Gino traveled solo to the continent to join him. Arriving in Boulogne two hours before his father could get there to meet him, he was happily chatting away in a sidewalk café when Daddy arrived.

With his father, Gino toured the ruined landscape of the Somme and the Aisne, at age eleven helping soldiers salvage valuables from the mud. The horror of the war only dimly pierced the "very nice time" he wrote his grandmother that he was having. "There are not any houses standing in the village here. I have seen a lot of broken-up towns. They are much worse than I thought they would be. . . . The other day Daddy and I went to a grotto. There were lots of caves and rocks there and a great many bombs and bullets and helmets and machine-gun bands lying about, it was a great pity all the caves were spoilt by shells."

After the first year's delay, Gino was enrolled in Bexhill Academy, a boarding school near Newcastle upon Tyne in the far north. During his four years there, he proved to be a mediocre student and showed little interest in sports, but left a mark for two odd achievements: he swam the length of a forty-foot pool underwater, and he built a crystal wireless set functional enough to pick up Morse code signals from ships in the English Channel.

After Bexhill, at age twelve, Gino applied to the Royal Navy (a common practice for schoolboys at the time). He passed the entrance interview but failed his exams. No doubt his father had dreamed of a military career for his son, but no path in life would have been more constraining to his antic spirit. Gino himself expressed no disappointment at failing to pursue the life of a midshipman.

The navy setback took a perverse toll, however. The Colonel had set his heart on Gino's attending Eton, his own school, but the failed exams derailed Gino's acceptance at that elite academy. Instead he would attend Lancing College, a second-rate public-school situated near the dreary south coast of West Sussex. The school was renowned for its grim Christian austerity and distrust of any education smacking of the progressive.

Very little testimony as to how Gino fared during his five and a half years at Lancing has trickled down, but Jamie Scott, who met Gino at Cambridge, got to know him better than anyone else, and later wrote his biography, insisted, "Lancing, with its stern, bracing life, gave him the health he lacked and made his body almost as strong as his will."

A fellow student during his Lancing years was Evelyn Waugh (though there is no evidence the boys were friends), who left a searing portrait of school life in his mordant memoir, *A Little Learning.* Waugh had been slated to go up to Sherborne, but after his older brother Alec wrote a scandalous roman à clef about his own years at the school, Evelyn became persona non grata there. "Lancing was monastic, indeed, and mediaeval in the full sense of the English Gothic revival," Waugh wrote. The school diet was appalling: "The food in Hall would have provoked mutiny in a mid-Victorian poor-house and it grew steadily worse." Waugh might have taken a clue of what was to come from the first remarks he heard out of the mouths of boys returning for their second year: "O God. Same old House Room. Same old smell."

Dress and behavior were rigidly codified: "Costume was entirely subfusc for the first two years; then coloured socks were permissible; in the Sixth Form coloured ties. For the first year hands must be kept from the trouser pockets, for the second year they could be inserted but with the jacket raised, not drawn back." The yards themselves were strictly regulated: "Grass, in which the grounds abounded, was in general forbidden territory; every plot was the preserve of some privileged caste, the most sacred being the Lower Quad, where only school prefects might tread."

Beatings with canes were carried out routinely, dispassionately, by everyone from the housemaster to upperclassmen, with only the meagerest explanation as to why they had been earned. In general, though "friendlessness was at first inevitable . . . , odium was personal and something quite new to me."

Waugh survived Lancing by ducking his head, conforming, and saving his satiric wrath for adulthood. Gino Watkins, though, seems to have taken a different tack. On the thin testimony of a couple of classmates, he seems to have floated through on disdain and quixotic rebellion. Recalled fellow student Robert Lea: "We used to spend our time laughing at each other, laughing at the absurdities of our subjects and of those who taught us, always magnifying the slightest chance of a joke. Gino was continually laughing at all around him, especially at himself. . . . He never discussed anything seriously."

According to Jamie Scott, who was a Cambridge blue in rugby and a strong cricketeer, at Lancing Gino "was no more serious at games."

> Football [soccer] was cold and pointless. He did not want to score a goal any more than he wanted to be kicked in the shins. "And," he said, "you get so cold rushing through the air like this." The only possible way to get amusement out of such a sport was by shouting, "Pass, pass!" until he got the ball and then kicking it into the dyke which ran beside the field. Cricket was a waste of pleasant weather.

Yet Gino found joy and accomplishment in swimming and cross-country running, representing Lancing in meets, and in learning to shoot a rifle in the school's officers' training corps. Even here, his rebellious streak prevailed, as when, on an away outing for a shooting match, he took his crew to a public house to get rowdy drunk. At Lancing Gino also discovered a passion for climbing the school buildings. "[S]ome of the things he did were almost too spectacular to watch," Robert Lea testified. "I remember on another occasion he climbed round the Masters Tower, an incredibly difficult business; I have never heard of anyone else who tried it."

During the Lancing years, on holidays Gino often joined his father for tramping and hunting excursions, one of the best a winter jaunt through southern Ireland. Uneasy feelings over Irish versus English rule were ironed into bonhomie, as the gamekeeper told Gino and his father, "It only cost four shillings to get drunk under the English and now it costs a pound—glory be to God, Sir, we'd loik the English back." That evening in the inn, dancing sprang up. The worn-out father went to bed, only to be wakened by his seventeen-year-old son: "The ball's broken out again and the band is drunk. It's splendid fun—you must come." During the Cambridge years, dancing became one of Gino's great passions.

After the war, Henry George Watkins Sr. lingered as often and as long as he could in France and Switzerland, hunting, hiking, and spending his fortune. The family, no longer able to afford Eaton Place, moved its ménage to Onslow Crescent, in a less toney London district. Tension with Jennie about raising the three children went mostly unspoken, even as it worked its harm. When he was home, she wrote her husband, "Gino is a darling, much quieter than he used to be," and "I love the children to like home and each other best."

Gino dearly loved his mother, Scott later remembered. But the life he yearned for lay far beyond London and Lancing. In the summer of 1923 Colonel Watkins brought the whole family, including Nanny Dennis, over to France for six weeks of rest and play out of Chamonix. At first the family sampled only the tame tourist excursions to the foot of the Mer de Glace, the Mauvais Pas, and the Glacier des Bossons. In the evenings sixteen-year-old Gino danced with the hotel gigolo, but soon he was slipping away unannounced, to be discovered by his father ensconced in the guides' hut listening to their tales of derring-do and begging them for itineraries.

One day he told his father that the ascent of the Aiguille de l'M, the graceful tower that stares over Chamonix from the south, was an easy climb. A guide led the way, roped to Gino, who was roped to his father. Halfway up, the Colonel lost it. In his own account, he was stymied in "a thing called a pillar box out of which it had seemed impossible to move." Gino, who had seconded the pitch without any trouble, called out, "Let yourself swing out on the rope, Daddy." As the trembling, exhausted man, with a hearty pull from the guide, floundered onto the belay ledge, Gino chirped, "Oh, Daddy, I'm afraid you aren't really enjoying yourself!"

Gino couldn't get enough of this new sport, mountaineering. Yet in an odd way, that enthusiasm had nothing to do with exploring. As a compositional prodigy, Mozart had been steeped in, obsessed by, music since early childhood. By the age of twelve he had written eight symphonies and his first opera. But Watkins the exploratory prodigy emerged only at age twenty. Robert Lea, remembering Gino at Lancing laughing at everything, added, "He never spoke about his future work. Personally I doubt if he had any intention of exploring." A couple of years later, Gino sat down with his father to discuss his career prospects. As the Colonel reported, "[T]hat night after dinner for the first time he expressed his love of the out-door life and his hope of an outdoor profession. Farming, army life, Kenya and Canada were discussed, but exploration as a profession entered neither of our heads."

Back at Lancing after Chamonix, Gino discovered a master named E. B. Gordon who was keen to learn climbing. The next spring the two drove up to the Lake District for a week based in the Wastwater Hotel in Wasdale Head. With another novice, they managed to get up climbs on the Great Gable and Scafell. According to Gordon, "Gino always led. He never did anything sensational for the sake of his reputation; he used to get very tired; he hated getting cold, but he never gave up anything which he had started."

In the summer of 1925, after graduating from Lancing, Gino made his way back to Chamonix without his father. Gordon joined him briefly, but after the master left, Gino found other ropemates among the locals, including an ambitious young Frenchman determined to win his guide's certificate. Among his deeds, Gino ticked off the Aiguille de l'M and the Grands Charmoz, the latter in deep snow. "None of the guides at the Montenvers will believe we did it," he bragged in a letter home.

On the Charmoz, Gino suffered his first close calls. His account in the same letter mixes fear with thrill:

Coming down a great stone came bounding down a couloir in which we were: it just missed my head and flicked my arm. I thought it had broken it to begin with, as I could hardly move it, but it is much better now but very stiff. Coming down the glacier near the bottom where we were unroped, my leg went through into a crevasse, which was very frightening.

Near the end of summer, on the verge of returning to England, Gino extended his holiday when the Colonel came over to hunt chamois with his son in the Austrian Tirol. The local jaeger, or hunter, was joined by a mountain guide. For several days the foursome ranged the crags and cirques, and Gino killed his first chamois with a single shot.

Father and son dined on roast chamois that evening in the hut, and were off again early the next morning. The guide led up a steep couloir, bypassing snow-filled crevasses, on terrain tricky enough that, as the Colonel would report, "Often the man in front had to give a helping hand to the man behind him."

At the top of the couloir, the guide set off to traverse an eighteen-inch ledge above a high cliff. Gino followed, rifle and rucksack on his back, as his father and the jaeger dubiously crept across the exposed passageway.

To steady himself, Gino reached above his head and clasped a protruding rock. It came loose in his hand, and he fell backward off the cliff, plunging, bouncing, and scraping to a halt 150 feet below. A thin band of snow had snagged his body just above another precipice. Knocked out cold, Gino later said he had felt no pain during the fall.

On the ledge, the three men stood paralyzed, unable to spot Gino's body below. Realizing the situation, Gino sent up a feeble cry, "I'm all right!"

During the fifteen minutes it took the guide and the jaeger to climb carefully down to the victim (leaving the Colonel marooned on the ledge), Gino assessed his injuries. He knew he was badly wounded, but he grew impatient. Spotting a chamois perched on an outcrop far above, he retrieved his rifle, braced himself in the snow, and fired. The shot missed.

The rescue down to the mountain hut took three hours. When the Colonel first saw his son, he was being carried on the guide's back with the jaeger steadying them both. Blood was gushing freely from Gino's head, and he later admitted he was "in excruciating pain." There followed, across several days, a carry 3,700 feet down the valley with Gino immobilized on a stretcher made of two poles and a blanket; an overnight in the inn with a doctor wielding chloroform and sewing stitches; and a final trundle down to the highest village in a wheelbarrow.

Father and son would never again share a ramble in the outdoors. Three weeks later Gino went up to Cambridge. For the rest of his life, he parted his hair to the right side, to cover the scar on his left from the long fall in the Tirol.

* * *

At the age of twenty, Gino Watkins sounds like the kind of daredevil who learns little or nothing from his near misses. Indeed, he seems almost to revel in them. Of course most young men who take crazy risks think they're immortal, but one wonders whether for Gino the accident in the Tirol served as some kind of release of the rage or joy of the anarchic baby who had wadded up bread and butter and thrown it at the ceiling. In Greenland three years later, Gino would unnerve his teammates by claiming that in climbing, "It was the thrill of being frightened that I most enjoyed."

If so, that psychic bent looms as the antithesis of what you normally want in an expedition leader. In the makeup of great explorers such as Nansen, Amundsen, or Shackleton, for all their boldness, there is no germ of the daredevil or thrill-seeker.

At Trinity College, during the first months, Gino was forbidden by his doctor to open a book because of the concussion from his fall. Instead, he at once applied to join the Cambridge University Air Squadron, the first recruit for the first civil flying school established in Britain.

At the end of his apprenticeship, belying the indifferent student he had been at Bexhill and Lancing, he won top marks in a field of ninety.

Flying in small planes partook for Gino of the vagabond freedom of climbing. Yet in his second year, he suffered another close call. On a training flight, as he practiced aerial photography and navigation from the copilot's seat, the plane lost its engine. The pilot glided groundward, hoping for a soft landing, but struck the earth hard and flipped the plane upside down. Having not bothered to fix his seat belt, Gino was flung from the plane, while the pilot was suspended upside down in the cockpit. Running up, Gino cheerily inquired, "You all right, sir?" before freeing the man, then persuading him to pose for a photo atop the wreckage, like a great hunter with his kill.

One more thrill sought and found, one more devil dared?

Neglecting his studies through his first year at Cambridge, Gino escaped with flying and excursions up north to rock-climb. In the summer of 1926 Gino returned to Chamonix, hooked up again with the ambitious French climber angling to become a guide, and knocked off an impressive roster of some twenty guideless ascents and traverses, including the Aiguilles des Droites, Courtes, and Moine, the traverse of the Grépon, and a long traverse of Monte Rosa.

At Cambridge Gino further indulged his appetite for building climbing. There had long been a tradition at the university, with its crenellated and decorative Gothic and Renaissance architecture mingling with idiosyncrasies in Georgian brick, of illicit forays by undergraduates, usually at night. A decade after Watkins's university tenure, a now-legendary guidebook called *The Night Climbers of Cambridge*, by one Whipplesnaith, sang the deeds of dozens of outlaw ascensionists, all of them nameless but well photographed, complete with precise route descriptions, friendly disputes over difficulty, and tips for avoiding arrest.

Gino's building climbs, however, seem to have been performed solo, outside of any coterie of like-minded conspirators. And they reached beyond Cambridge. When a friend remarked on the inspiring sight of Salisbury Cathedral, 140 miles southwest of the university, Gino recounted his assault: "Yes, that's just what I thought as soon as I saw it. . . . I slipped off in the dark. I started climbing the spire in gym shoes, and was going quite well when it started to rain and the stones became impossibly slippery. I had quite a job getting down."

Gino's passions for flying and climbing may have sprung from his wanderlust, but they did not point him toward exploration. The routes he tackled on Scafell or the Chamonix aiguilles were classic test pieces. Over Christmas 1926 Gino joined his Lancing friends Quintin Riley and the master E. B. Gordon at the ski resort of Arosa in eastern Switzerland. New to the game, Gino took to skiing as hungrily as he had to flying and climbing. By the fourth day of instruction, he was launched onto the expert slopes.

Three weeks later, on returning to Cambridge, he found the prospect of a second term so dreary that he made a snap decision. A friend greeted him outside Trinity: "Hallo, Watkins, you've been to Switzerland."

"Yes, and I'm just going back."

"But the term's started: you can't."

"Why not?"

Gino recruited part of his family to join him for several weeks of winter sport and dancing in the Arosa inn. Not content with skiing, Gino also tried to master the luge. Speed itself was the drug. One evening he took a run when the slope was officially closed, and as he hurtled through the twilight at some thirty-five miles an hour, he collided head-on with a horse-drawn sleigh. As he careened off the luge, one shaft of it pierced his thigh. Laid up for two weeks with the serious wound, he hung out in the inn. "I must say he bears these things most wonderfully," his father wrote home. "Luckily he doesn't mind noise, as until we can get him upstairs he is between a telephone and a jazz band, which he likes." Soon he was back on skis, perfecting the art of falling only on the side of his good leg.

During that outing at Arosa, Gino turned twenty. Within the year, Jamie Scott, his Cambridge classmate, would meet him, and Scott has left a vivid description of the man who would so change the course of his own life:

> He wore a double-breasted blue suit and his fair hair, parted on the right side, swept smoothly back above a high unwrinkled forehead. His face was thin in flesh but not narrow, for the bones on either side ran straight down from the temples till they suddenly turned inwards and slightly toward the angle of his jaw. Looking at his mouth one noticed chiefly a line of very white teeth. His eyes as he bent over his papers were shaded by long lashes, like a girl's, but when he looked up one saw, as one had expected from his fair coloration, that they were close and blue. They were lively eyes which mirrored his thoughts before his lips could frame them. . . . They were the only feature he could not control. . . .

Perhaps the best photograph of Watkins in his early twenties captures him in a pinstriped three-piece suit with a neatly knotted tie closing a point collar. His left hand is tucked into his pants pocket, while his right, resting on crossed knees, holds a cigarette, even though Gino seldom smoked. He had developed the affectation of carrying a rolled umbrella everywhere, even on sunny days. Slight of build, he is evidently handsome and exudes confidence, though his pose projects the faint suggestion of a dandy.

Feckless student, flamboyant climber and skier, thrill-seeker, lover of dancing, jazz, and parties—this vignette seems irreconcilable with the man who would become one of the great Arctic explorers of the twentieth century. The single event that transformed Gino Watkins was the lecture by Raymond Priestley that he attended his first Cambridge year. We have only a vague idea what Priestley spoke about that evening, but no polar veteran had more extraordinary deeds under his belt.

Because of the intense antipathy that grew between Scott and Shackleton on Scott's first quest for the South Pole (1901–04), only a single man signed on as a member both of Shackleton's *Nimrod* expedition (1907–09) and Scott's fatal *Terra Nova* exploit (1910–13). That man was Raymond Priestley, and he remains more than a century later one of the most underappreciated of all the British explorers of Antarctica. (Priestley, in fact, asked Shackleton for permission to join Scott in 1910, but after grudgingly acquiescing, "the Boss" treated his former teammate as a traitor for the rest of his life.)

On the *Nimrod* expedition, Priestley was the key man enlisted with the thankless job of laying depots along the Beardmore Glacier to support Shackleton's final thrust, which met defeat only 112 miles short of the South Pole. Three years later, Priestley was in charge of Scott's six-man Northern Party, exploring Victoria Land west of the team's base while Scott and his four doomed companions pushed through to the pole. When the expedition ship was blocked by ice from picking up Priestley's team after its eight-week stint, the men dug a cramped snow cave into a drift and wintered over for seven months, barely subsisting on seal and penguin. As soon as spring came in October 1912, the men, crippled by enteritis, hobbled for five weeks, dragging their sledges along the coast, and barely made it back to the base at Cape Evans.

That magnificent survival feat was utterly eclipsed by the slowly unfolding tragedy of Scott's polar party, as they weakened and perished on their desperate trek back from the pole. In Apsley Cherry-Garrard's masterly *The Worst Journey in the World*, the Northern Party's epic goes all but unmentioned. In 1914 Priestley wrote his own account, in a vivid memoir called *Antarctic Adventure*. The book is virtually unread today.

The talk that Gino and Quintin Riley attended that evening in 1925 was titled "Man in the Polar Regions." After Shackleton's catastrophic failure and survival tour de force on the *Endurance* expedition from 1914 to 1917, Antarctic exploration had lapsed into a limbo for more than a decade. No doubt Priestley's talk was full of hints instead for great discoveries to be made in the Arctic. Gino was not shy about approaching the veteran, and Priestley, taking a liking to the young man, introduced him to James Wordie, a tutor at St. John's College in Cambridge. Wordie had been one of the twenty-two men left on Elephant Island after the sinking of the *Endurance*, while Shackleton, Frank Worsley, and four others performed the now-legendary open-boat journey and traverse of South Georgia Island that saved the whole team.

Despite that harrowing ordeal, Wordie was still keen on exploring the polar regions a decade later. Gino learned, in fact, that the tutor was leading a small team to probe the east coast of Greenland in the summer of 1926, only months away. Unfortunately, the roster for that exploit (which included August Courtauld) was full; but Wordie promised Gino a berth on a kindred expedition slated for 1927.

At once Gino started training for that grand campaign, making long runs around the quads and through the streets of Cambridge, devising a rigorous exercise program, and sleeping with an open window and only a light sheet covering him on the coldest nights to toughen himself up for the Arctic. All the climbing in the Lakes and the Alps, the manic nighttime ascents of university buildings, even the full-bore skiing holidays, became part of his training. Meanwhile, he devoured the rich literature of Arctic exploration dating all the way back to the first European attempts to force the Northwest Passage, starting with John Cabot in 1497.

And he reformed his slacker's approach to studying. According to Jamie Scott, at the end of his freshman year, as exams loomed, Gino hired a "crammer" to fill his head with all the knowledge from the books in a certain course that Gino had never opened all term. As Gino sat nonchalantly in the back of the room, paying only half-attention, imagining that he could somehow absorb the learning by osmosis, the crammer exploded: "I've got no time for you, Watkins. I'm far too busy with people who have some chance of passing this exam."

Scott: "The effect was dynamite. He walked back to his room and took out the books which he had bought but never opened. He had only a fortnight in which to do a year's work, but he had found his inspiration." Gino passed the course with first-class marks.

From his truant flight to Switzerland to perfect his skiing, Gino returned to Cambridge in the spring of 1927, excited to join the expedition to Greenland the coming summer. But James Wordie had bad news. For various reasons, finances among them, the veteran explorer had been unable to pull the trip together. He would have to postpone the expedition, he darkly confessed, for at least a year, maybe two, maybe even longer.

At age twenty, with no expedition experience of his own more daring than a couple of trips to tourist-thronged Chamonix, almost any budding explorer would have bitten his knuckles and accepted fate. Gino chose a different course.

Jamie Scott, writing in 1935 after sharing two major expeditions with his best friend, glides past the extraordinary challenge Gino set for himself in the spring of 1927 as if it were merely par for the Watkins course: "Gino's reaction was original, but natural enough to himself. He wanted to visit the Arctic: there was nobody to take him: therefore he would lead an expedition of his own."

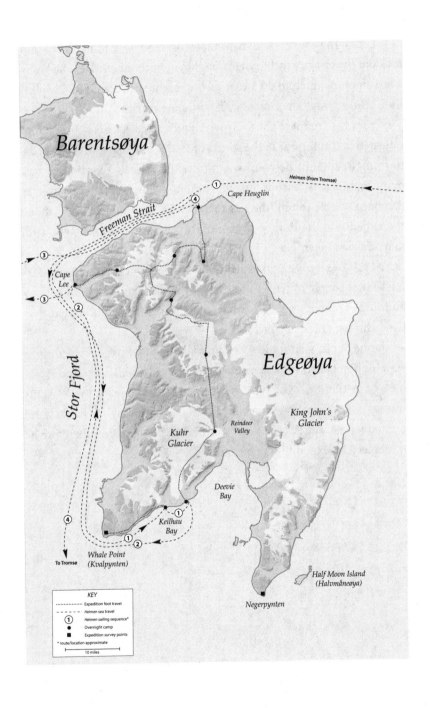

Barentsøya

Heimen (from Tromsø)

① Cape Heuglin

④

Freeman Strait

③

Cape
Lee

②

③

Stor Fjord

Edgeøya

King John's
Glacier

Kuhr
Glacier

Reindeer
Valley

Deevie
Bay

④

① Keilhau
Bay

②

Whale Point
(Kvalpynten)

To Tromsø

Half Moon Island
(Halvmåneøya)

Negerpynten

KEY
.... Expedition foot travel
‒ ‒ ‒ ‒ Heimen sea travel
① Heimen sailing sequence*
● Overnight camp
■ Expedition survey points
* route/location approximate

10 miles

TWO

Edgeøya

In choosing his own objective for an expedition, Gino followed Wordie's lead. In 1919 and 1920 his Cambridge mentor had co-led small ventures to Svalbard, an archipelago of four major islands situated well to the north of Norway, between latitudes 76 and 81. Never prehistorically inhabited, Svalbard was visited in 1596 by the Dutch explorer Willem Barents, who had sailed blind into the high latitudes in search of that elusive shortcut to China, the Northeast Passage. Because the waters off Svalbard were incredibly rich in whales, Barents's discovery unleashed a frenzy, lasting more than three centuries, of ruthless killing of the great ocean mammals, as well as the slaughter of walruses, seals, and polar bears. Entrepreneurs from Russia, the Netherlands, Britain, Norway, Sweden, and America descended on Svalbard in a free-for-all of marine exploitation.

Nearly all that activity took place around Spitsbergen, the largest of the four islands, whose shores are bathed by the Gulf Stream, which surges north off its west coast. The east coast of Svalbard, on the other hand, is battered by a relentless flow of polar ice pack pushing south, rendering the other three islands—Nordaustlandet, Barentsøya, and Edgeøya—inaccessible by ship for eleven months of the year, open only during a window from mid-August into early September. Some years that window never opens.

Gino decided to explore Edgeøya, the southeast corner of the archipelago. The island, seventy miles long by forty-five wide, had been discovered in 1616 by a shadowy merchant sailor from Lancashire named Thomas Edge. But through the centuries, thanks to its difficulty of access, Edgeøya had been neglected almost entirely by the hunters of whales and walruses.

A single Russo-Swedish team from 1899 to 1901 had accomplished a perfunctory survey of Edgeøya's west coast. The rest of the island remained unmapped and unknown. No party had yet pushed into the interior, thought to be covered by a sprawling ice cap.

Yet in the spring of 1927, Gino faced a formidable challenge. In three months, he needed to raise funds, lease a ship, buy and transport tons of gear and food, and recruit a team for his expedition. Before that season, Gino had scarcely organized anything more complicated than a family ski outing to Switzerland. He was not a joiner, much less a leader. With his pals Quintin Riley and Robert Lea, he had laughed his way through dreary Lancing. His first year at Cambridge had been marked by distractions, sudden escapes, and desultory scholarship.

That Gino had a gift not only for friendship but for persuasion would come to light during the next five years of his short life. Yet it remains a mystery how, during those frantic months of preparation, he put together a solid team of eight companions. Seven of them were Cambridge undergraduates or young alumni. The wild card in the crew was Henry Morshead, forty-four years old that spring, a longtime official with the Survey of India with extensive Himalayan experience. In 1921, on the first Everest expedition, Morshead had helped George Mallory pioneer the route to the elusive North Col, the key to all attempts from the Tibetan side for the next thirty years. Back on Everest the next year, Morshead reached 25,000 feet with Mallory, E. F. Norton, and Howard Somervell, a new record. He was generally acclaimed as one of the Himalayas' great explorers: of Morshead, Norton swore, "I never met a harder man."

But on that 1922 thrust, inadequately outfitted, Morshead suffered crippling frostbite of toes and hands, later having three fingers amputated. What convinced the forty-four-year-old veteran to join a twenty-year-old upstart on a junket to Svalbard five years later—perhaps Wordie's intercession?—we can only speculate. But on Edgeøya, Morshead became Gino's staunchest partner.

In a frenzy of fund-raising, Gino won pledges from a Cambridge fund and the Royal Geographical Society totaling £250, or $10,600 in 2022 American dollars. It was enough to lease a two-masted ship, the *Heimen*, and hire an old-fashioned Norwegian skipper named Lars Jacobsen. The nine-day voyage out of Tromsø featured the dubious sport of shooting polar bears from the ship as they lolled on ice floes, bouts of seasickness that laid almost all hands low (Gino would suffer exorbitantly from the malady throughout his life), and a storm that tore away the ship's wireless aerial and flooded the belowdecks. Captain Jacobsen disdained the ship's navigational instruments, preferring instead to climb to the crow's nest and sail by dead reckoning. When Edgeøya at last came into sight, he was chagrined to find the ship not on the island's west coast, but well to the east of it, amid loose ice pack.

To solve that glitch, he brilliantly piloted the *Heimen* through the seldom-traversed Freeman Strait between Barentsøya and Edgeøya, then down the west coast of the island. On July 31 the team made landfall at Keilhau Bay on the south coast, not far from where the Russo-Swedish team had launched its survey.

As soon as he set foot on land, Gino was afire with the urge to discover, charging off solo to climb the nearest high hill, topping out 2,000 feet above the sea. From that summit he gazed east to size up the unexplored interior. He was acutely aware of how little time the expedition had. The *Heimen* was scheduled to pick the men up at Cape Lee on the northwest corner of the island on August 20, barely three weeks hence. So Gino made the first order of business traversing Edgeøya from west to east, surveying as the men proceeded, then returning by a different route to cover as much new ground as possible. Ninety miles all told, which Gino hoped to cover in eight or ten days.

On August 1 he chose four men to join him on the traverse. In the time-honored tradition of Scott and Shackleton, they would man-haul a sledge with all their gear and rations packed on top of it.

From the start, the trek could hardly have gone worse. Gino had reconnoitered the first several miles the day before, realizing at once that sledging would be impossible until the team reached permanent ice. So on August 1 the five men backpacked all their baggage, dragging and carrying the empty sledge, until they got to a glacier that spilled from the central ice cap. Yet even there, the sledging was nasty, on bare ice riven with glacial streams. The day had started fine and clear, but by evening a heavy fog descended. Only partway up the glacier, the team stopped to camp.

The fog stayed thick and low to the ground for the next ten days. Morale plunged with each wasted day. To save weight, the men had brought along no books. A couple of them filled the silences with singing. Years later, Gino could recite on harmonica the South American ditties that team member V. S. Forbes had warbled inside the tent.

During fugitive clearings, the men moved their camp a hundred yards or so, and climbed to nearby summits to survey with plane table, only "to sit shivering and straining their eyes for the points they could not see." Despite the maddening halt the fog had carved out of the team's tight schedule, the five men apparently got along well with one another. And this raises for the first time one of the deepest puzzles built into the genius of Gino Watkins. How did he lead?

How did a twenty-year-old on his first expedition mold the efforts of eight companions, all older than himself, one with decades of experience in the Himalayas, into a coherent team? Shackleton and Scott wielded the authority of their leadership through force of example, but also through institutional hierarchy. Trained in the Royal Navy, Scott so rigidly enforced his discipline that throughout both his long expeditions, the daily messes his teams sat down to on shipboard and in hut were segregated by "officers" and "men," with different meals.

Gino never gave orders as Scott, Shackleton, or even the more egalitarian Amundsen and Nansen, had. Yet on all four of his expeditions, not a single teammate ever sabotaged a plan, much less threatened mutiny. One insight into Gino's leadership style comes in a conundrum offered by Forbes after the Edgeøya journey. "The most extraordinary thing," he said, "was that Gino gave no orders in the ordinary sense and we all thought we were doing exactly what we wanted to do. But afterwards we realised that we had done precisely what he meant us to do."

After a couple of days, Gino went ahead with Morshead to try at least to reach the high point of the ice cap. They traveled light, leaving the sledge behind, carrying little food and no stove. But then they were stranded in a two-day blizzard.

They gave up on the traverse on August 11, and all five men straggled back to the southwest coast, reaching Keilhau Bay on the twelfth. The expedition was half over, and the team had accomplished almost nothing. Of the forty-five miles of wilderness stretching from west coast to east, the men, wrestling with their finicky sledge and humping loads in their rucksacks, had covered barely twelve.

Gino's response to the setback was to redouble the challenge: his men would now try to traverse Edgeøya the long way, seventy miles from south to north. With only two companions, carrying only a small tent and eight days of food, ditching the sledge, he set out on August 13. The rest of the party, under Morshead's leadership, would sail north on the *Heimen* and meet the traversers at Cape Lee.

The trio quickly regained the glacier, but on the ice cap itself they were stymied by fog and twelve-inch-deep slush. After a poor first day's travel, though, they hit their stride, averaging twenty miles a day. In the middle of the ice cap, they made a startling discovery. Crossing a glacial col, they saw ahead of them not more unbroken ice cap, but an open, fertile valley, "an oasis of quite luxuriant vegetation among the barren glaciers and snow domes which rose on three sides of it." In one stroke the trio had discovered that the center of Edgeøya was not sheathed in uniform ice cap, but rather in patches and lobes of thick ice interspersed with small open basins. (Even today, geographers struggle to explain the oddities of Edgeøya's interior.)

Descending into the oasis, the men stumbled upon a herd of reindeer. Animals that had never before encountered humans, they were slow to scare. Forbes tied a discarded pair of antlers to his head and crept within twenty yards of the reindeer before they skittered away, only to stop and stare back from fifty yards.

Curiously, though they had seen numerous polar bears on ice floes on the cruise between Tromsø and Edgeøya, they encountered not a single one on their traverse. This despite the fact that, then as now, Edgeøya is a migratory corridor for polar bears, one of the most densely traveled in the Arctic.

On August 16 the men realized that they could easily finish the south-to-north traverse of the island in only five days, rather than the eight they had budgeted for it. But on that day, still twenty miles from Cape Lee, they heard voices. It was Morshead and a teammate, out surveying from their base at the cape, where the *Heimen* waited to pick up the team and sail for Norway. After the joyous reunion, Gino's trio joined their teammates in their camp only six miles inland.

There was no getting around it: the expedition was now an unqualified success. Three men had traversed Edgeøya the long way, discovering the essential features of the interior: an unknown upland that had thwarted exploration throughout the previous three centuries.

But the men got to talking in camp that night, and Gino's antic restlessness surfaced. Simply to head down to Cape Lee and call it a job well done seemed too easy. The lure of the unknown still raced in his blood, and after all the days of fog, the weather was holding glorious. To trudge down to Cape Lee, crossing land Morshead had already surveyed, seemed "a waste of time."

So that evening in camp Gino devised Plan B, a spontaneous extension of the journey just to eke out one more discovery. Instead of heading down to Cape Lee and the ship, Gino persuaded Morshead and Forbes to head northeast with him across yet more terra incognita, aiming for Cape Heuglin, the extreme northeast point on Edgeøya at the entrance to Freeman Strait. That jaunt would add another forty miles to the roster of exploration.

In his fits of manic ambition, Gino could be cavalier or even oblivious to the needs of others. As the men separated on the morning of August 17, Gino told the other two to return to the ship and instruct Captain Jacobsen to load up and head back east through Freeman Strait. The rendezvous at Cape Heuglin would take place, Gino decreed, by midnight on August 21. One can imagine Jacobsen's annoyance at being ordered on this uncertain and perilous mission by a twenty-year-old in absentia.

Of course, as soon as Gino's trio started out on their new jaunt, the weather turned bad. The men advanced only a few miles the first day; the second, none at all, as they hunkered through a storm in their cramped tent. Suddenly five days to cover forty miles looked like anything but a piece of cake. But with the mist clearing on the morning of August 19, Gino pushed his partners hard.

During the next few days, winter seemed to arrive early. The first night snow blew inside the tent, soaking the men's sleeping bags. Paraffin leaked into the food in Forbes's rucksack, ruining the biscuits. August 21 dawned so thick with fog that the men despaired of finding their way to Cape Heuglin. But since the ship was due at midnight, Gino decided to force the blind march down to the coast.

On that stagger north, Morshead collapsed. Forbes took much of the man's load (Gino's already being the heaviest), and the younger men tried to rally the veteran. "I was very afraid he would never get there," wrote Gino in his diary, "as at every halt he just dropped down looking absolutely dead, and he could only get to his feet again with assistance." Out of the miasma, the shape of a hut gradually cohered. The men rushed on, imagining the crew holed up in it and the ship anchored nearby. But the hut was ruined, full of old bones and bits of parched skin left by some long-ago hunters, and the *Heimen* was nowhere in sight.

Midnight came and went. The men were almost out of food. During brief clearings of the fog, the men stared at the empty ocean. Even the sanguine Gino Watkins grew alarmed. In his diary, he wrote,

> I really don't know what has happened to [the ship]. I realize one thing for which I am to blame and that is that I am sure one ought always to carry a rifle, even if one has already got very heavy packs. If the *Heimen* has run aground and can't get to us, if we had rifles we might manage to shoot enough reindeer and freeze them in to live through the winter. As it is we shall be bound to starve if she does not come. It is very disappointing.

All through August 22, the three exhausted men peered through the fog and waited, unable to get warm in their soaked clothes. In early afternoon, Gino and Forbes started to hike west along the coast, leaving Morshead in the tent by the ruined hut. And then, out of the murk, the vague shape of the *Heimen* coalesced.

That evening the famished trio gorged on roast goose, from a bird a teammate on the ship had shot the same day. Belowdecks the exhausted trio changed clothes and got warm. Captain Jacobsen turned the ship around and headed back west through Freeman Strait.

On August 30, the men spent their last night on the *Heimen*, docked in the Tromsø harbor. "We all sang songs," reported Gino. "It was a very merry evening."

*   *   *

Back at Cambridge for his third term, Gino once again neglected his studies as he feverishly put his notes in order, developed photographs, and wrote up the data his "scientists" had compiled, all in preparation for the lecture he would give at the Royal Geographic Society in February.

By the fall of 1927, Gino's family was reduced to a straitened existence. In the house on Onslow Crescent, Gino's mother struggled with her finances, as Gino's sister and younger brother, still at home, completed the ménage. Gone were the servants, though the ever-loyal Nanny Dennis kept the household from falling apart.

About a year earlier, Gino's father had contracted tuberculosis. Having squandered most of the inheritance that had once supported the lavish life of Eaton Place, the Colonel chose not to return to England, but to install himself in a sanatorium in Davos, Switzerland, where he would while out his remaining days. He had, in effect, absconded from his own family.

On February 20, 1928, in the august Aeolian Hall at the RGS, Gino stood up before 500 fellows to deliver his account of the Edgeøya expedition. If he was nervous in front of such a gathering, he hid it well. His charm, his youthful good looks, his modesty, and a few flashes of his dry wit soon won over even the crustiest of the old walruses in attendance.

Gino's mother was in the audience. Afterward she wrote her husband to replay the momentous evening.

> I felt in a ghastly condition of nerves, and as the hall got fuller and fuller I got iller and iller! Amid much clapping . . . Gino stepped onto the platform. Gino looking about sixteen, very pale and quiet. . . .
>
> He began at once, very quietly and modestly and rather too quickly, but he gradually settled down to a very easy pleasant style and was quite excellent—made one or two jokes which were laughed at heartily. . . .

He received an ovation at the end and I can't tell you how many ripping things were said about him.

The talk had been a complete success. At the end of the ceremony, the RGS president, Sir Charles Close, announced that Gino had been elected as a fellow of the society. But there was a technical problem. He was too young to become a fellow. Then and there, the RGS amended its bylaws and voted Gino in. He thus became the youngest fellow in the ninety-eight-year history of the institution.

Teammates, friends, and Gino's mother repaired to supper at the Savoy. The dancing wound on until 2:00 a.m. The next morning, friend after friend rang up Mrs. Watkins "to say that you are a damned lucky woman." In the letter to Gino's father, she added, "I wish you could have been [there] last night."

By February Gino was deeply immersed in the planning for his next expedition. Already the RGS had agreed to sponsor it.

A few weeks after the celebration in Aeolian Hall, Gino's mother rose early and took a train to Eastbourne on the south coast. From the station she hired a taxi to drive her out to Beachy Head, the great chalk cliff overlooking the sea. Then she disappeared. It was presumed that she walked to the edge and jumped. Her body was never found.

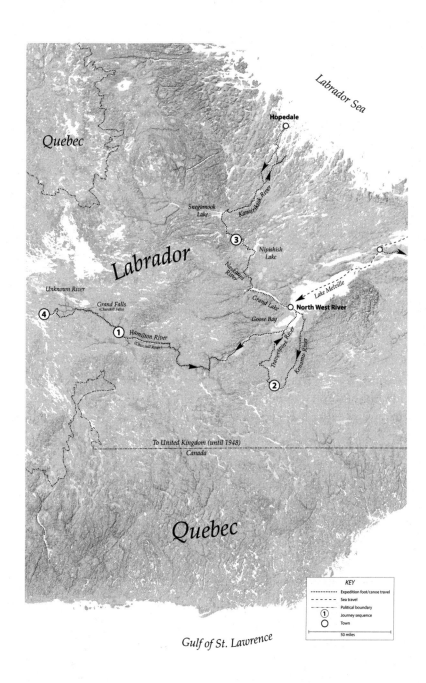

Quebec

Labrador Sea

Hopedale

Snegamook Lake

Kanairiktok River

Labrador

③

Nipishish Lake

Unknown River

Naskaupi River

Grand Falls
(Churchill Falls)

Grand Lake

Lake Melville

④

①

Hamilton River
(Churchill River)

North West River

Goose Bay

Traverspine River

Kenamu River

②

To United Kingdom (until 1948)

Canada

Quebec

Gulf of St. Lawrence

KEY

··········· Expedition foot/canoe travel

‒ ‒ ‒ Sea travel

‒·‒·‒ Political boundary

① Journey sequence

○ Town

50 miles

THREE

The Land That God Gave Cain

FOR GINO WATKINS, exploring Edgeøya was a drug, mainlined into a vein that had coursed before only with conventional blood. Even before he got back to Cambridge for the fall 1927 term, Gino was scheming up his next expedition. A man he met in Tromsø had recently been to Franz Josef Land, the Arctic archipelago much more remote than Svalbard. The spark was lit. Before Captain Jacobsen could escape, Gino booked the *Heimen* for the summer of 1928.

Within weeks, the pipe dream had assumed major proportions in Gino's head. As he wrote a friend:

> [The ship] will be large enough to carry an aeroplane. I shall try to map out the eastern end of the islands and also do one flight out eastwards, as I am sure there must be some land between Franz Josef Land and Siberia. If there is any land there it will be very important in a few years' time, as if an air station were established there, and an air route established via Franz Josef Land to Japan, the distance would be decreased by over 2,000 miles.

Nothing came of this wild plan. But there in Gino's brain, three years before he would realize it on the ground, lay the whole plan of the BAARE in embryo, with Tokyo as the endpoint rather than Winnipeg.

None of Gino's Edgeøya teammates was up for another expedition the following year. Morshead had dashed back to India, and thence to Burma, where he was soon promoted to Director of the Survey for that colony.

Even as he passed again through the gates of Trinity College, however, Gino was frantic to be elsewhere. Another exploratory goal he sketched out was even wilder than the Franz Josef fantasy. The jaunty phrases he uttered as he tried it out on friends almost invited ridicule:

> We would leave England in the autumn with a party of three and go by cargo boat (that's cheap and good enough) to Jibouti at the northern end of the Red Sea. One can get to Addis Ababa by train. There we would hire camels and ponies for the ride across the desert. We would carry flat-bottomed, collapsible boats made of metal like a seaplane float. When we reached the Blue Nile we would drift down it to Khartoum, surveying as we went. . . . I believe there are funny little men who live in caves and shoot poisoned arrows as one goes past. It ought to be rather amusing.

(The first descent of the Blue Nile would unfold only in 1968, by a seventy-man British army party in rubber rafts. One boatman drowned, and several rafts and quantities of cargo were lost. The team pulled out nowhere near Khartoum, but at Shafartak, still well within Ethiopia.)

Gino's extravagant blueprints for future expeditions, which he would share again and again with teammates in the midst of their present journeys, were like wild jazz riffs on straightforward melodies: Gino as the Bix Beiderbecke, so to speak, of terrestrial exploration.

In late autumn 1927, Gino was rescued from his exploratory daydreams by a solid proposal from the RGS, which promised financial backing. It was to make a thorough survey of the newly decreed border between Labrador and Quebec.

Sometime around AD 1000, Vikings had made the European discovery of North America. But that cardinal event was demonstrated on the ground only in 1960, when the brilliant archaeologists Helge and Anne Ingstad found the remains of a Viking settlement at L'Anse aux Meadows on the northern tip of Newfoundland. Meanwhile, in 1497, Captain John Cabot claimed the discovery of North America for Great Britain. During the next three centuries French mariners and voyageurs, including Jacques Cartier and Samuel de Champlain, made their own bold excursions up the St. Lawrence River and into the Great Lakes. The gigantic subcontinent that would become Canada teetered between French and British rule until the French and Indian War at the end of the eighteenth century ended with Britain victorious. Canada became a British colony.

Left in a kind of limbo were the island of Newfoundland and the adjoining wilds of Labrador to the north. The Crown loftily maintained that all that territory belonged solely to Great Britain, entirely separate from Canada. Only in 1902 did an aggrieved Canadian government insist that Labrador amounted merely to a strip of coastline extending no more than one mile inland. The rest, it argued, was Quebec.

Not so, responded the Crown: Labrador stretched all the way west from the shoreline to the height of land, the divide that separated waters flowing into the Atlantic from those that emptied into Hudson and Ungava bays and the Gulf of St. Lawrence. The trouble was that, despite journeys of exploration prosecuted by government officials and amateur adventurers from the 1830s on, nobody knew exactly where that height of land lay, and the sketch maps those voyagers brought back were shaky at best.

After a quarter century of wrangling between Canada and Great Britain, the Privy Council ruled in favor of Britain in 1927, about the time that Gino got back from Edgeøya. The height of land it was to be. And who better than the Royal Geographic Society's latest fair-haired boy explorer to go to Labrador, survey his way inland, and map the boundary divide once and for all?

Gino was galvanized by the project. At once he read everything he could get his hands on about Labrador. As he later remarked, "When I started the work I hardly knew where Labrador was. By the time I had finished it I had planned a year's expedition to that country."

The challenges of Svalbard and Labrador could hardly have been more different. Edgeøya was uninhabited, treeless, half-covered in ice cap. But Labrador had been the homeland for centuries to two closely related native peoples: the Montagnais and Naskapi Indians (collectively the Innu, as they called themselves) to the south, the Inuit or Eskimos along the northern coast. Since 1836, the Hudson's Bay Company had established trading posts all along the coast, with a few well inland up the great rivers. In addition, Moravian missionaries had succeeded in converting nearly all the Innu and Inuit to at least a nominal Christianity.

For sixty years before 1928, a remarkable band of trappers, some of them Métis (of mixed native and British ancestry), some Scots and English loners and fortune seekers, had blazed trails deep into the interior, built cabins (which they called "telts") at regular intervals, laid out their traps, and often wintered alone in the backcountry, returning only in late spring to sell their furs at the trading posts. Nobody knew interior Labrador better than these men—except the Innu and Inuit. Yet these were not the kinds of wanderers who made maps: all their knowledge, like that of the natives, was stored in their heads.

Gino recognized at once that he would have to learn a whole new way of exploring to tackle Labrador. Instead of fog-bound Arctic plateaus, he would need to master a limitless landscape of dense forest, traveling through the maze of rivers, lakes, and marshes that spangled the wilderness. It was a place where you had to climb a tree to figure out where you were. A place where you could get lost for good just fifteen minutes off the trail.

For Gino, it was all enticing. But instead of a nine-man team squeezed into a two-month summer window, he craved a deeper immersion. He would take only two companions, one of them for the first summer only. But he would spend a whole year in the Labrador interior. He looked forward to wintering over, and to learning a whole new set of exploratory skills.

It was at this point that Gino met the man who would become his best friend and biographer. James Maurice (Jamie) Scott was Gino's Cambridge classmate, a mere month older than his future companion, but by their junior year they had yet to meet. One day in the spring of 1928, Scott found himself with several other students in Raymond Priestley's suite, listening to the veteran tell Antarctic war stories, the kind too personal to be included in talks such as "Man in the Polar Regions." "With a heavy breakfast inside me," Scott recounted, "[I] said that I should like to travel about a bit before settling down to a job."

Without hesitation, Priestley answered, "Watkins is planning to go to Labrador. I'll write to him and you had better go and see him."

"I was very vague about Labrador and I had never heard of Watkins," Scott confessed. But some days later, with Priestley's further nudge, he approached the Trinity door lettered with "H. G. Watkins" across the top.

I knocked and was told to come in; then apologized and prepared to withdraw. The rather frail, elegant young man lounging in an armchair could not possibly be the leader of exploring parties, nor could his companion who was dressed more simply in open dressing gown and a wet bath towel.

I felt certain that I was in the wrong room. But the man in the arm-chair sprang up and called me back with "I'm Watkins, are you

Scott?" The man in the dressing gown disappeared and we began to talk.

Gino pulled out a map of Labrador and drew his finger along the journeys he intended to make. "Could I ski or walk in snowshoes?" he asked. "No," Scott answered. "Well that was a pity but one could quickly learn." In a breathless monologue, Gino extolled the landscape, emphasizing the "virtually unexplored" height of land that formed the boundary he had been charged to map. Travel would be by canoe in summer, by dog sledge in winter. Most of the wilderness "was unknown except from the reports of trappers and Indians." Summing up the adventure to come, Gino promised that "There was, in fact, much to be discovered at the expense of a lot of energy and a spicing of danger."

Thus Gino worked his infectious charm. As Scott later put it, "I had come to Watkins' room in a spirit of vague curiosity. I walked home half an hour later with nothing in my mind but Labrador and him."

Shortly thereafter Gino recruited his second teammate, the man who would stay for the first summer only. He was a friend named Lionel Leslie, seven years Gino's senior, an army veteran who had traveled in Africa and Burma. Through the spring months of 1928, the three men feverishly prepared for the expedition.

Then, suddenly, Gino's mother disappeared near the cliff edge of Beachy Head.

Gino was utterly devastated. With his father off in Switzerland, he was thrust into the role of head of his fractured family, the protector of his sister, Pam, and brother, Toby. They needed him more than he needed Labrador. Still in shock, he decided to call off the whole exploit to which he had devoted nearly every waking thought for half a year.

At this point, Gino's father intervened. The old sportsman trumped the grieving husband. He wrote to Gino and argued forcefully that he not give up the Labrador expedition. He offered to return to England to look after Pam and Toby during Gino's absence. The proposal swayed Gino, then won him over: Labrador was back on the drawing board.

* * *

By July 19, 1928, the three men were installed at North West River, a Hudson Bay trading post sixty miles inland across massive Lake Melville and the hub for all exploratory journeys into the interior, as well as the launching pad for the trappers who maintained their lines and trails hundreds of miles west up the great rivers. At latitude 54 degrees north, the village lies well to the south of the Arctic Circle. But all of Labrador might as well be part of the Arctic, so cold are its winters, so heavy its snows.

Always impatient, Gino was eager to be off on the team's first exploratory journey, but all kinds of things had to be arranged first. On Edgeøya, he had learned the hard way the sheer futility of man-hauling sledges. In Labrador he was determined to hire dogs and teach himself how to sledge with them, but at first none of the locals had any huskies to spare. Accustomed though he was to setting off on his own, Gino realized that he also needed to hire an experienced trapper to serve as the party's guide, for those hardened loners not only knew the country well but also knew how to survive it.

Stymied by the unavailability of sledge dogs, Gino made the rounds of the assembled trappers in search of a man willing to guide the expedition throughout the coming year, which would mean giving up his own pursuit of fur-bearing animals through the prime winter months. He offered each candidate a salary of one British pound per day. That may sound measly, but in 1928 it was a decent wage: translated into today's dollars, the equivalent of $84. Yet at first Gino found no takers. "The good men felt that their traps would yield more than the pound a day which we offered," wrote Scott, "while nothing would persuade the less enterprising trappers to go so far from home."

The whole enterprise of Labrador trapping was infused with the bonanza mentality of gold-rush fever. The trappers lived by the legends of fabulous fortunes won overnight, such as John Grove's luck on an unnamed lake deep in the wilderness, where he caught so many foxes that he made $7,000 in a single season. (In today's dollars, an unthinkable $86,000.)

At last Gino snagged the attention of a seasoned trapper named Robert Michelin, whose father, brothers, and cousins all pursued the lonely and dangerous business of the backcountry quest for marten, fox, weasel, ermine, lynx, mink, muskrat, beaver, otter, wolf, and wolverine. Though gruff and a man of few words, Michelin at once won the trust of all three Englishmen: as Scott wryly noted, "He might even see our point of view to the extent of joining in uncomfortable and pointless adventures."

On the journey up the coast earlier in the month, Gino had recorded in his diary the men's first glimpse of Labrador: "At last—our first sight of 'the land that God gave Cain.' It looks rather rugged and bleak and reminds one of the northern coast of Norway."

In 1534 the French explorer Jacques Cartier had sailed west into the Gulf of St. Lawrence, the first European to penetrate that great basin. He took one look at the terrain stretching north on the starboard side and was appalled. In his *Première Relation*, he wrote, "I am rather inclined to believe that this is the land that God gave to Cain." (In Genesis 4, after asking Cain where his brother Abel was, only to hear the impudent answer, "Am I my brother's keeper?" God laid his punishment on the fratricide: "And now thou art cursed from the earth. . . . When thou tillest the ground, it shall not henceforth yield unto thee her strength; a fugitive and a vagabond shalt thou be in the earth.")

The barren shore that Cartier beheld is actually part of Quebec today. But ever since the French navigator bestowed his flippant comment on that northland, The Land That God Gave Cain has served as an epithet for Labrador.

In 1933, four years after the expedition ended, Jamie Scott published a book-length narrative of the Labrador adventure. He chose for his title *The Land That God Gave Cain.* The book is one of the master-pieces of North American exploration, page after page of vivid descriptions of the country, rich in the evocation of its characters, and governed throughout by a dry wit that nonetheless dodges the pitfalls of pallid understatement. Sadly, like so much having to do with Gino Watkins, the book is virtually unknown and unread today.

As soon as they had arrived at North West River, Watkins, Scott, and Leslie realized how impossible their mission was. (The savvier of Gino's RGS benefactors might also have suspected as much.) Almost completely unsurveyed, the height of land between Labrador and Quebec was not some clean ridgeline above timberline visible from miles away, but an indistinct divide smothered in dense forests (and, farther north, in the swampy furrows of trackless taiga and tundra).

Modern cartographers have proved that the Quebec-Labrador border wanders through all its zigs and zags for 2,170 miles, making it the longest interprovincial boundary in Canada—longer even than the boundary between Alberta and British Columbia, provinces whose sizes, respectively, are more than twice and three times as big as Labrador.

Instead of trying to determine the divide throughout the territory, Gino resolved to make three deep thrusts into the little-known wilderness, surveying and mapping as he went, following the rivers that the trappers themselves had used as dangerous highways into the game-rich interior. The first journey would occupy the autumn of 1928; the second, the darkest, coldest months of winter; the third, the rampaging winter-spring transition of 1929. Each voyage, under the tutelage of Robert Michelin, would school the Englishmen in a whole new set of wilderness skills.

All three journeys had their triumphs and perils, and all three would result in brilliantly detailed maps, full of practical hints for future travelers. But among the three, it was the middle journey, through the heart of winter, that saw Gino Watkins emerge as a great explorer and survivalist in the mold of Nansen rather than of Shackleton.

Gino's first exploratory junket was an ambitious loop far to the south of the "base camp" of North West River: up the relatively well-traveled Kenamu River and down the little-known Traverspine. To connect the two, the men would make a blind portage across a bewildering chain of lakes, a traverse unknown to the trappers. Gino hired a second guide, Douglas Best, who had built a telt a hundred miles up the Kenamu.

All travel was by canoe. Ascending the Kenamu, the three acolytes learned the hard way from their guides the several techniques of upstream canoeing, from paddling on calm water to poling, lining, wading chest deep and pulling the boats, and, when the rapids were impassable, portaging. Gino had hoped to reach the height of land overlooking Quebec, but so rigorous was the going, they never got within fifty miles of it.

The men had packed supplies for more than two months, but thanks to soaking and spillage, they eventually ran out of almost all their essentials, including tobacco. More than once, the dreamy Lionel Leslie got lost off the trail and had to be found with shouts and tree-climbing.

On the traverse to the Traverspine, Gino came into his own, navigating brilliantly by Polaris and Arcturus. But when the Traverspine abruptly plunged into a deep gorge with nearly vertical walls, the men had to resort to a ten-mile portage that took six grueling days.

Gino seemed completely caught up in his new adventure. But privately, he was haunted by the suicide of his mother. His unresolved grief emerged in an uncharacteristic homesickness. In one diary entry he wrote:

> We are camped in a soaking wet wood trying to dry some of our things in front of a fire, but it is raining too hard. It is all rather depressing. I am just going to crawl into a soaking wet sleeping-bag.

I would give anything to be at home and have Nanny to tuck me up
in a nice warm bed and bring a cup of Ovaltine.

The difficult journey ended up taking seven weeks, from August 20
through October 1. One of the finest passages in *The Land That God
Gave Cain* records the return of the men, including the habitually abste-
mious guides, to "civilization" in the form of the first furnished telt.

Everyone knows that it is foolish to give alcohol to men who are not
used to it: but one cannot relax, celebrate, and be reasonable all at the
same time, and for weeks we had been looking forward to this party.
Watkins compromised by producing only enough whisky to illuminate a Christmas pudding. But next morning, as we surveyed the
result—two broken windows, an overturned stove and a bite in the
leg—we realized that one cannot be too careful. A trapper once told
me that he could get drunk on pickles; it is a tall story, but I believe it.

* * *

The morning after the return from the Traverspine, Lionel Leslie
boarded the mail ship as he parted from Scott and Watkins, his summer's spree in Labrador behind him. Until the last minute, Gino feverishly scrawled letters to send home. The anguish over his mother's
unfathomable departure from the living, as well as his worry about the
family left in England, had never been far from his thoughts on the
Kenamu and the Traverspine. The tug of home clashed with his exploratory ambition as it never would again in his short life. To his loved ones
he sent that untruthful platitude that all adventurers toss off as they set
out into the unknown: "Remember we are not going to do anything in
the least dangerous."

Before their winter journey could begin, Gino determined to use the
down time to learn everything he could about the country. The three
men had been lent a small house on the "Indian bank" in North West

River, and with their gramophone as the magic lure, they put on records, opened the door, and waited for visitors to arrive. As Scott related,

> To the Indians the gramophone was irresistible, especially when they discovered that they were allowed to play it for themselves. They could not understand a word that was sung, but that seemed to be an advantage rather than an objection, since it prevented them from getting bored by any one song and enabled them to go on and on with the same record without having to bother about changing it. In return we expected our visitors to tell us all they knew.

To their surprise, the men learned that the Indians were better at reading and drawing maps than the trappers. The only glitch was distance. The natives thought in terms of a day's carry, not in terms of miles. Enlisting a trapper who spoke some Naskapi as a translator, Gino spent hours with his informants. "Yet whether they talked wisely or wildly," Scott observed, "Watkins listened without a sign of boredom, remembered what he wanted and forgot the rest." In this way the team learned all kinds of tantalizing lore: about a clearing where one hunter had slain many caribou and commemorated the spot by tying a set of antlers to a tree (Gino & co. later found that very tree); about a place nearby where a whole family had starved to death. And as the men earned the Indians' trust, they were told about the mysterious Snegamook Lake, which no white man had ever visited. At once Snegamook became a top priority for Gino.

The trappers, likewise seduced by the gramophone, imparted their own lore. Reticent at first, "they soon opened their hearts to a visitor from Outside who . . . treated everyone as socially equal and mentally superior to himself." The veteran hunters tended to know their own traplines better than the country beyond, but, as Scott remarked, "they understood more clearly than the Indians the sort of things we wanted to know." The trappers' wives were happy to make deerskin mittens, sealskin boots, and slippers to wear as inserts for the still-tenderfooted Englishmen.

Late autumn was the dead season for all the Labrador trappers. They needed to wait for the first snows and the freeze-up on lakes and rivers before they could launch the winter journeys that made or broke each year's fortune in furs. The hardiest of those men traveled hundreds of miles solo into the wilderness and stayed out until March.

Both on Edgeøya and in Labrador, Gino had been keen to accomplish real science. Mapping the height-of-land divide was the raison d'être of the year-long expedition. In his proposal, he had even waxed enthusiastic about Labrador's commercial promise, envisioning today's eco-nightmare of forests leveled for timber and a hydroelectric plant on Grand Falls (realized in the 1960s by the Churchill Falls dam and power plant) with undiluted zeal. But another side of Gino, perhaps the truer side, burst out occasionally in mad soliloquies like the one he startled Scott with during one of the down days at North West River:

> You know [Gino said], some day I'm going to run a really good expedition. There won't be any rot about taking nothing but scientists and dull people like that. We'll have Art Fowler to play the ukelele and Jack Hylton's band and a complete chorus. We'll take the *Mauretania* to Franz Josef Land and build a really good house and have a frightfully good time. It would be grand, for all the conventional geographers would be so shocked, and actually, of course, we'd get a tremendous lot done; for people work so much better when they're happy.

That last phrase could stand as a formula for Watkins's leadership style.

For the winter months, Gino had planned a far more ambitious trek than the late-summer loop: a push up Grand Lake and the Naskapi River, in quest of Snegamook Lake; thence overland out of the forests and across the taiga to Hopedale, an Inuit village on the northern coast.

Then back again, all the way home to North West River. Robert Michelin had agreed to hire on once more as guide.

The waiting drove Gino crazy. Michelin lived not in North West River, but with his wife and numerous children in a cabin up the Hamilton River, near the mouth of the Traverspine. After the first big snowfall on November 7, Gino expected the man to arrive by canoe any minute to help him and Scott prepare for the big winter journey. But for days there was no sign of or word from the trapper.

The uncertainty drove Gino to a rash decision, which ended up plunging Scott and himself into the closest scrape of the whole Labrador year. If Michelin would not come to North West River, he and Scott would travel up the Hamilton to find out what was delaying their guide. Gigantic Lake Melville, which separated North West River from the coast, seldom froze over, but its southern lobe, Goose Bay, now began to form a skin of ice. Across that unstable surface, Gino and an initially reluctant Scott would trek to roust out Michelin in his own den.

The canny Hudson Bay factor in North West River, one Mr. Thavnet, horrified by Gino's plan, begged the men not to go. Instead Gino chose to travel light—no sledge, just two men, one axe, their lunch, and their snowshoes. If the ice was thin, a light tread was the answer. They set out on the morning of November 18.

Halfway across Goose Bay, the tiny oasis of Rabbit Island served as a standard rest stop. Having found no trouble thus far, Scott and Watkins gobbled some chocolate and headed on. At once the conditions changed, for across the southern half of Goose Bay, a current dumped its waters into the minimally lower Lake Melville. Gino took the lead five yards ahead of Scott, testing the ice by probing with the all-purpose axe. If it took two or more blows to pierce the frozen surface, the men deemed the going safe.

Suddenly the snow cover vanished: only a thin layer of new ice lay between the men and the frigid moving water underneath. "Over this we traveled with short sliding steps, never lifting our feet," Scott would later write, "so that our weight might be more evenly distributed and there should be no sudden shock." Briefly they debated putting on their snowshoes, but decided not to, for fear that they might stumble and fall.

They skittered on. Now the ice ahead started cracking with loud reports, as masses of of it split off from other masses. Gino held the axe at shoulder height and let it fall. The blade instantly punched a hole in the fragile skin, and a fountain of water gushed out. "All around we could hear the water gurgling and muttering," Scott recalled. "It was a thought that if one went through one would be swept downstream and would come up again some distance below the hole."

The men decided to turn back. But no sooner had they started toward Rabbit Island, two miles away, than the ice cracked open across their outward tracks. They were in effect marooned on an ice floe in a moving river, a floe that might crumble at any moment. Scott and Gino tried to find a continuation above and below their track, but everywhere new cracks appeared. "We had to think moving," wrote Scott, "to stand still would have been fatal."

Somehow Gino signaled that they might as well go ahead as go back. Abandoning any pretense of testing the ice, they shuffled forward, "crouching low that we might not fall . . . , feeling [the ice] sink down and sway like a stretched tarpaulin beneath our weight."

Five minutes later Gino dared take another swing with the axe. The surface had abruptly deepened to a safe four inches. Scott:

> We laughed and stood still to look behind us. The events of life which one describes rather loosely as "pleasant" may be divided under two heads: those which one enjoys while they are happening and those which are pleasing chiefly in retrospect. What we had done was definitely not of the former variety, but though I disliked the operation so intensely while it lasted I cannot avoid classing it as a pleasure

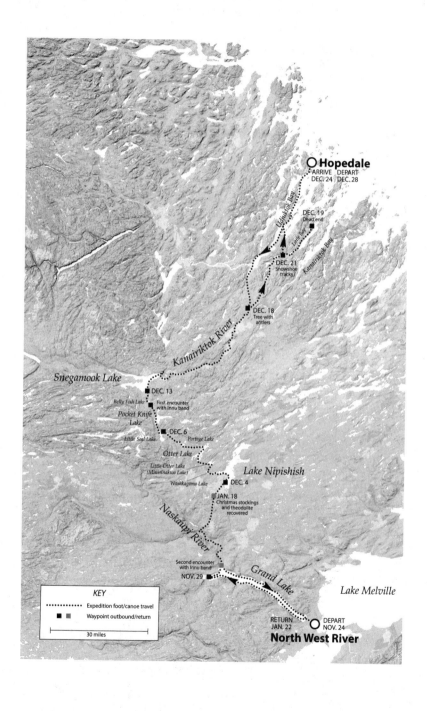

O **Hopedale**
ARRIVE   DEPART
DEC. 24   DEC. 28

DEC. 19
Dead end

Ujutok Bay

Kanairiktok Bay

Little Bay

DEC. 21
Snowshoe
tracks

DEC. 18
Tree with
antlers

Kanairiktok River

Snegamook Lake

DEC. 13

Belly Fish Lake   First encounter
                      with Innu band
*Pocket Knife
Lake*

DEC. 6

Little Seal Lake   *Portage Lake*

Otter Lake

*Little Otter Lake
(Minuttinakwa Lake)*   *Lake Nipishish*

Wishkagama Lake   DEC. 4

JAN. 18
Christmas stockings
and theodolite
recovered

Naskaupi River

Second encounter
with Innu band

NOV. 29   *Grand Lake*

                                    Lake Melville

KEY

••••••• Expedition foot/canoe travel

■ ■ Waypoint outbound/return

———————
30 miles

RETURN   O DEPART
JAN. 22   NOV. 24
North West River

because of the wonderful feeling of satisfaction which it gave rise to
as soon as it was safely over.

A day and a half later, Scott and Watkins arrived at Robert Michelin's cabin. There they discovered the simple reason for the guide's
apparent procrastination: he had stepped on a rusty nail and had just
recovered. The seasoned trapper had none of the hectic urgency that
had pushed Gino and Scott to their reckless thin-ice traverse of Goose
Bay. Two days later, he led his charges back to North West River by a
safer route on solider ice. "Our friends seemed pleased but a little surprised," Scott reported, "for they had not expected to see us, they said,
till the ice thawed in the spring and our bodies were washed up somewhere down the bay."

The grand winter excursion up the Naskapi River and across to
Hopedale began on November 24. Around North West River Scott and
Gino borrowed or leased five huskies for the Hopedale trek. With his
usual dry wit, Scott characterized the foibles and virtues of each animal.

Robert had his own hunting-dog, a highly intelligent white male
called Buntin. Gino had Sneider, a fat, very friendly ball of soft
brown fur with appealing eyes and secret vices, and Frisk of the
euphemistic name, a melancholy, russet-coloured cabhorse of a dog
who hauled very hard because he had never done anything else in
his life. I had Beauty, a lovable but rather too sex-conscious bitch,
and Chub, jet black from head to tail, a temperamental hauler but a
great fighter who bit me when we first met and became my greatest
canine friend.

Both men granted that on this longer second jaunt, they would be
novices in the art of Labrador winter travel, schooled by a taciturn master. Scott captures Michelin's pedagogical style:

Robert taught us to recognize the tracks of the various fur-bearing animals and from him we picked up such varied knowledge as how to set traps and dry skins; how to hunt, cook and eat partridges [as the Labrador hunters called ptarmigan]; how to mend snowshoes and moccasins; how to light a stove and make a brushwood bed; how to keep an axe sharp and use it rightly; in fact, how to be comfortable. It needed a certain amount of initiative to acquire this knowledge, for Robert rarely told us anything. In the course of a year I don't believe he ever said that I was doing something the wrong way. He did things one way and if you did them another it was your look-out.

North West River lies at the junction of two of Labrador's greatest rivers: the Hamilton, with its headwaters far to the west beyond Grand Falls, and the Naskapi, which winds southeast from a jumbled chaos of small lakes and brooks, terrain that even the trappers knew far less well than they did the Hamilton. For Gino the great appeal was semi-mythic Snegamook Lake, known only to the Innu. Around the gramophone in late summer, despite the language barrier, he had pumped the Indians for details. They had drawn him a crude map in pencil.

The first fifty-five miles of the Naskapi above North West River was not river at all, but long, skinny Grand Lake. At the end of November, not yet frozen over, it presented a problem. With the help of two volunteers, a borrowed rowboat, and a canoe, the team carried their Nansen sledge, two light Indian "toboggans," food and fuel for four weeks, a tent and a stove, and the five unwilling dogs (which had to be tossed bodily on board), halfway up Grand Lake. Not to mention the five men piloting the boats.

From there on, Gino and Scott had to sledge along the banks of the lake, weaving through trees and thickets. Michelin charged ahead in the canoe, leaving Gino straggling behind, and Scott, saddled with the Nansen sledge, "a very bad last." Now Scott got his first lesson in husky ferocity, as he carelessly dropped his deerskin gloves on the snow to perform some delicate adjustment, only to have Sneider scarf them down like an after-dinner snack.

For the sledging novices, that first week on Grand Lake was an ordeal of misery and frustration. Even as they struggled on, the men shot game that came their way to supplement their provisions. In this fashion, Gino and Scott learned how to catch and cook a porcupine, the handiest prey of all. "Porcupines," Scott insisted, "are very good to eat and delightfully easy to kill. They are slow-moving, stupid and make practically no attempt at protective colouring." Because they are voracious consumers of spruce bark, the animals leave tell-tale rings of bare tree, "a mark which not even the most inexperienced hunter could miss." Once those signs are spotted, "you need only look around until you discover the porcupine. Either it is in a hole under some upturned root, or else it may be quietly feeding above your head and can be brought down with a shot or by merely shaking the tree."

To cook a porcupine, the hunter must "swinge" it. "The art lies in burning off the quills without singeing the flesh. . . . If the job is done carelessly the savoury meal is spoilt by a taste which resembles nothing so much as the smell of burning feathers."

On November 29, the men reached the head of Grand Lake. Several streams emptied into the lake, and it was not easy to discern which was the Naskapi. Michelin found the right current, but after following the Naskapi some ten miles northwest, he veered off to ascend the much smaller Crooked River, which flowed directly south thirty miles from sprawling Lake Nipishish. Along this route thirty years before, Michelin's father, a legend among the trappers, had cleared and blazed a veritable highway, and built four telts within the thirty miles, some of them equipped with real glass windows. Robert had inherited the knowledge of the route, but "in the meantime small saplings had poached upon the road, and though they were no serious hindrance to a well-controlled hand-sledge they formed a succession of tempting maypoles round which a team of dogs might wind their traces."

As was his wont, Michelin charged off in the lead, man-hauling a small toboggan, leaving his clients to flounder with their snowshoes, the dogs, and the Nansen sledge. At least once Gino and Scott gave up short of the next telt and pitched their tent instead. At first on gaining the plateau above the Naskapi valley, they found better going, which tempted them to wild optimism, as they dared predict a trip to Hopedale in two weeks and a return to North West River by Christmas. But then their floundering inefficiency kicked in, and they were able to cover only five miles a day.

That discouragement in turn dictated a decision that well might have had dire consequences. To lighten their loads, Gino and Scott cached the Christmas stockings their friends at North West River had given them. That meant disappointment; but they also cached "all the clothes we can spare," Scott's gun and books, and the precious theodolite, without which they could not accurately determine latitude. The sacrifices left their loads eighty pounds lighter, which made a crucial difference.

On December 4 the three men finally reached the last telt, on the shore of Lake Nipishish. Meanwhile Gino had developed a sore throat that left his voice raspy and thin, and a fever over 100 degrees Fahrenheit. Uncomplaining, he put in a brutal day of jogging at six miles an hour alongside the sledge to reach Nipishish. But he needed a rest, and Sunday (observed as a day of rest by all the trappers) couldn't come too soon.

Again and again, Scott would report that he only got to know the real Gino Watkins during those down days. Lying in his sleeping bag, Gino would deliver pronunciamentos without obvious provocation. In this way Scott heard his friend declare that "a man's private life was his own and one must not judge him by it so long as he kept it safely within bounds." And out of nowhere, Gino would deliver dogmatic screeds:

A person ought to have complete control of himself in every way. He ought to be able to sleep as easily with the window shut as with the window open. To me the Englishman who can't get into a railway carriage without throwing open the window is every bit as bad as the Frenchman who can't get into a railway carriage without shutting the window.

(What was this fetish about control about?)

Man ought to be a slave to nothing [he went on]. There is nothing that he ought not to be able to do without, be it tobacco, drink, or women. Of all the vices I think drink is the worst, for a drunk man loses control and I think there is nothing more degrading than that.

Scott found these dicta puzzling, almost off-putting, even while he scrutinized them for clues to his friend's inner life. It bears remembering that such utterances came from the mouth of a twenty-one-year-old, and that young men are the most inclined to lay down laws by which the world ought to run.

Beyond Nipishish, the country was unknown to Michelin. For the first time Gino and Scott felt that they were exploring new land, terra incognita except to the Indians. And here, the penciled map Gino had been given by the Naskapi took over from Michelin's seat-of-the-pants knowledge of the wilderness.

Somewhere off to the west lay Little Seal Lake, but the intervening country was a maze of crooked brooks, tiny ponds, marshes, and woods charred by summer forest fires. It took the men ten days to cover thirty miles as the crow flies, but many more as they were forced to bend their path to the quirks of the terrain.

Even during these shortest days of the year, like all trappers Michelin called a break around noon for a "boil," as he gathered "a raft of large green logs to prevent [the stove] from melting its way down into the snow." All this for a mug of midday tea! "Watkins and I were often doubtful whether it repayed us for our trouble," Scott complained. But Michelin refused to drink cold water, insisting "that he knew of an Indian who had died from drinking of a frozen brook and of another who had collapsed and had to be carried into camp."

The camping routine each night was a major operation unto itself. The three-man tent came with no ridgepole, so each evening Michelin had to fell a stout spruce limb to supply the need.

On December 15 the men reached Little Seal Lake, following the unerring pencil line on the Naskapi map. According to the same document, mythic Snegamook Lake lay another fifteen miles due north, connected by streams the men could not mistake. Halfway there they reached a smaller pond whose Indian name translated as Pocket Knife Lake. It formed a kind of watershed, for it flowed south to Little Seal and north to Snegamook. It was also the highest point of land the men had yet reached in Labrador, as well as the farthest west.

Five miles north of Pocket Knife Lake, the three men ran into a band of Naskapi Indians fishing through holes in the ice on a small, round lake they called Big Belly Fish. Gino was overjoyed to see that some of them were the same men who had given him directions to Snegamook and had drawn the map. But at once he realized that this roving band of Innu was in trouble: the caribou had mostly failed to appear, and they were on the verge of starving. The band, in fact, was making their arduous way back to North West River in hopes of relief from the Hudson's Bay Company store.

Without hesitating, Gino gave the Indians a major part of their own rations, even though the three men were already worrying about whether they had enough food to reach Hopedale. The Innu also begged for tobacco, which Gino likewise delivered: no sacrifice for himself, only an occasional smoker, but a real deprivation for Scott and Michelin.

Those men even had to lend the supplicants their pipes, "for their own pipes they had cut up and chewed for the sake of the burning flavour which the wood retained."

In return, the Innu repaired the men's sledges and snowshoes, which had become dilapidated. And they assured Gino & co. that Hopedale lay only five or six days' travel away to the northeast. They drew another map indicating the river to follow and a crucial junction downstream where their path should diverge from the river.

The next day, as the Innu moved on south, Scott and Gino marveled at how even the women and children hauled their own sledges.

On December 14, Gino and Scott "cursed and floundered" their way up a series of hills smothered in soft powder snow. Scott lost his recording pencil, backtracked for his knife after it fell out of its sheath, and had to deal with Chub, who had chewed himself free from his traces. "In fact I had just decided that sledging was the world's worst torture," he later wrote, "when I reached the top of the [next] hill and found the other two gazing, like Keats' Cortez, at Lake Snegamook."

It was really only another lake, sixteen miles long by six wide, "but it had acquired for us a certain, perhaps rather sentimental, significance. It was the natural turning-point of our journey and, in many ways, its chief objective."

The surface of Snegamook was as smooth as a skating rink. The men took off their snowshoes and glided across it with an easy freedom they had not tasted since the trip began. The next day they found the outlet, which fed a river the coastal Inuit had named Canairiktok.

Generously but perhaps rashly, Gino had taken his Innu informants' word that Hopedale lay only five or six days away, and had given the hungry nomads all the food his own team would need only if the final lap of the journey stretched beyond six days. And the first three days of "exhilarating travel" on the Canairiktok seemed to validate that calculation. The river ran free and open, with a sledge-friendly pan of ice along its banks. The men averaged twenty miles a day.

The Naskapi informants had told Gino to pass three short water-falls, then look for a sign they had constructed to indicate where to leave the river. For the Canairiktok flowed not to Hopedale but to an inlet three fjords south of the small village. That sign—fugitive enough—was a single pair of deer antlers tied to a tree on the left bank.

The fourth day was a Sunday, but Michelin overcame his Sabbath scruples and set out to look for the antler landmark. It would not be until Tuesday, though, that he found the crucial marker. Climbing out of the river valley, the men suddenly confronted a puzzle that Scott called "the most difficult decision of the trip." Not one, but three trails led off from the bench above the Canairiktok, heading north, northeast, and east. Gino deliberated long and hard before choosing the middle path. Neither companion disputed his choice.

During those sixty miles of river travel, the men had left behind the dense forest to emerge on the northern Labrador barren grounds, where only patches of scrub forest relieved the taiga. Here Robert Michelin was out of his element. He could find no spruce limbs big enough to supply a ridgepole, so the tent had to be droopily pitched with guy lines anchored by stones. The men's provisions were drawing low, and they had no certainty that they were on the Hopedale trail.

It was not that the guide was unaccustomed to risk and disaster. Every trapper could name a half dozen peers who had drowned in rivers or starved to death, usually at the end of their winter excursions as they dashed back to North West River. Robert's own half brother John, with partner Sam Goudie, had headed home once after a very successful trapping season. In their impatience they stayed in their canoe as the rapids grew gnarly. Scott retells their story:

> They came around a corner and, just in front of them, the river swirled under a ledge of still unbroken ice. Goudie, who was kneeling in the bow, managed to jump on to the ice, but John Michelin had forced his legs under the tarpaulin which was lashed above the load

and could not move in time. Goudie caught him by his long hair and jerked him to safety as the canoe disappeared beneath them with all their food and two thousand dollars' worth of furs. They got home without much hardship [!], but they were poor men until the next season was over.

On December 19, their fifth day on the trail out of Snegamook, the men crested a hill and saw ahead of them a level expanse of white at the bottom of a valley. They hoped it was an estuary of the sea. Even if Hopedale did not appear at its far end, they reasoned that they could follow the coast or cross the frozen fjords to the north and reach the safety of the settlement. But Gino's anxiety crept into his diary: "If it is only a lake we are in a bad way. No dog food, a little pemmican for ourselves and thirteen biscuits each. I had to feed the dogs on our biscuits to-day."

The next morning the men discovered that the white expanse was not even a lake, but "a rotten little brook, open most of the way." More small hills interspersed with patches of stream followed. Driven by desperation, the trio pushed on. Then Scott broke through the ice and got soaked to his waist. But as he changed clothes, he swore that he could smell the sea in his soggy clothes. Pushing on, they came by late afternoon to the unmistakable inner crook of a fjord, confirmed by a tidecrack in the frozen surface.

"We hurried on expecting every moment to see the smoke of a house or some Eskimo driving home with his dogs after a day's hunting," Scott wrote. "But for six miles nothing interesting happened. . . . I have always hated bathos, but this was the worst anti-climax of my experience."

So they had arrived at the wrong fjord. But even out of food, the men thought they should be able to follow the coast north to Hopedale—wherever it was. Yet as soon as they set themselves on that course, they recognized that it was hopeless. The ice in the fjord was too thin to trust, and the cliffs plunging into it too steep to climb with sledges. They camped that night on the shore, eating their pemmican in the dark because they had burnt all their candles.

As was his wont, Gino now recorded in his diary a point-by-point analysis of the team's predicament. It included some grim assessments:

2. We do not know where the nearest habitation is. . . .

6. We have no dog food left and the dogs are very hungry.

7. We have hardly any food left ourselves.

As a last-ditch resort, the men pondered killing and eating the dogs one by one. But the overriding problem remained: where was Hopedale? Could it even lie *south* of the dead-end fjord? Even the equanimous Michelin was now reduced to something like panic. "If it snows we'll be bitched properly," he told his teammates.

A single alternative loomed. The day before, at the head of the fjord, the men had noticed a single old snowshoe track coming in from the left. Sure that houses and Inuit villagers lay just ahead, they had ignored the track. Now they must reverse their course and find the marks in the snow again—if only drifting snow had not covered them for good.

By the end of the next day, they had found the snowshoe track. Famished and exhausted, they set up the tent and tried to sleep. But Michelin prowled around and discovered a still-baited otter trap. Pushing on northward, lighting his way in the dark with matches, he made a thrilling discovery. Back at the tent he crowed to Scott and Gino that he had intersected the fresh trail of a komatik, an Inuit sledge, driven, he could detect, by no fewer than ten dogs.

December 23 was a Sunday, but no Sabbath taboo could keep the guide idle this close to Hopedale. Three hours into their pursuit the men came to the shore of another fjord. There the sledge track vanished into hard ice. Slowly the men recognized that Hopedale must be on the other side of the fjord. They stared and stared, frantic to find the distant shape of a square house, but every dark form on the far shore resolved into an inert boulder—until suddenly the men saw a puff of smoke rise from one of those blobs.

Within minutes the travelers were greeted by yapping dogs, children, and "an old woman who knew no word of English but understood at once that we were hungry." Two "red-cheeked girls" arrived with a sledge-load of fresh seal, on which Gino, Scott, and Michelin immediately gorged. The house—the farthest outpost of the Hopedale community—was the bailiwick of old Peter Tutu, who spoke some English. The whole extended family was gearing up to travel in style to Hopedale for Christmas Eve. The men in the Tutu household were bemused to learn "that we were travelling through the country not for hunting but merely to make a map."

Gino and Scott had initially hoped to get back to North West River by Christmas. Hopedale was instead a happy alternative. There the men were fêted for several days around Noël by a congregation of Inuit families mixed in with a Moravian missionary family and two Hudson's Bay Company agents. On December 25, as Gino wrote in his diary, "We all had an excellent Christmas lunch, with goose and mince pies. It is awfully nice, very comfortable and homelike."

But the piercing joy of escape and survival had suffused the three men two days before Christmas, as they rode the Inuit komatiks on the way from Peter Tutu's house to Hopedale, in a procession led by a dog team hauling a freshly felled Christmas tree. As Scott evoked that ride, "We followed, yodelling and singing carols, restless upon the sledge. The air was pricking our cheeks, nipping our ears, and our minds were drunk with our good fortune."

* * *

Although it was Robert Michelin who had performed the critical act of discovering the Inuit sledge team's recent track, it was Gino who provided the moral equilibrium to the trio during their harrowing days on the wrong fjord. From this episode Scott gleaned what was perhaps the keystone of his friend's leadership style. Yet a paradox lay at the very heart of that style, which bore no resemblance to the calming grace under pressure that leaders such as Amundsen and Shackleton exuded. It was a style that balanced rash daring with dogged competence, as might be expected in a man so young who assumed the leader's role so effortlessly. On the BAARE almost two year's hence, it would steer the fourteen-man team through one predicament and misfortune after another.

As Scott put his finger on that style,

> [Gino] could plan the most hair-raising journeys in which he seemed to deny the possibility of unforeseen delays; but once started he was pessimistic enough to accept every accident without complaint. . . . He explained that without optimism we could never plan any long journey, but that, in carrying it out, to be pessimistic was the next best thing to being forewarned as a defense against adversity.

The men expected the return journey to North West River to be a piece of cake. They knew the route by now, and after a month of toil, Scott and Gino had become expert dog sledgers themselves. But nothing would prove easy on the men's dash back to North West River. Constant new snowstorms dictated slow, heavy sledging in the forest. And Gino developed a nasty case of dysentery that lingered throughout the return, as he had to stop often on the trail to deliver a stream of diarrhea. As usual, he never complained.

Two happy events punctuated the tedium of the retreat. On the Naskapi River the men ran into the same band of Innu who had been close to starvation when met on the outward trek. In the meantime they had reached North West River and traded their furs, and now they were back in their homeland with more food than they could devour—much of which they shared with the three white men. And at the last telt south of Lake Nipishish they recovered the theodolite and the Christmas stockings they had cached to save weight a month earlier. Scott recounted the men's delight:

> We found our Christmas stockings and opened them, more excited than children. There were toys and sweets and books and onions; and the stockings themselves were useful. The onions, we heard later, had been put in as a joke, but now the frost had killed their strength and they had become as sweet as oranges. They mixed badly with cream chocolates, but on the whole we made an excellent meal. Afterwards I read aloud from *Boys' Book of Adventure*, all about Eskimos and icebergs, while Watkins and Robert did puzzles and played with dolls.

The men finally reached North West River on January 19. The journey they had expected to complete in less than four weeks had taken seven. Michelin had dashed ahead on the last home stretch, and the report he must have given of Gino's weakened state was dire enough that Mr. Thavnet, the Hudson's Bay Company factor, roused the nurse to admit the invalid at once to the mission hospital. "Watkins was bundled into bed with a glass of milk and a thermometer while I was settled in a comfortable chair before a more substantial meal and told that I must move about as little as possible," Scott recalled. "Left to ourselves we shook hands and then burst out laughing."

In his vulnerable moments, after six months in Labrador with two long, dangerous journeys under his belt, Gino might have been temperamentally ready to fold his cards and go home. But opposing the tug of England and family was the stronger compulsion to finish the job, to complete the program he had pledged to the Royal Geographical Society. In the mission hospital, on January 29 he had turned twenty-two. Now, in fact, he was keen to get back on the trail.

* * *

The third of Scott and Watkins's three major thrusts into the Labrador interior can be briefly summarized. It lasted from February 2 to March 31, during which the three men covered more than 600 miles. Most of that travel was on the Hamilton, Labrador's greatest river, which by 1929 was a well-mapped thoroughfare. Gino's goal, beyond the Hamilton, was the mysterious region around the romantically named Unknown River in the far west of the territory. And if possible, Gino would try at last to reach the height of land and stare into Quebec.

In the early 1920s, two small parties of explorers had briefly intersected Unknown River and barely glimpsed through the mists a previously unreported waterfall. But the sketch maps they produced were so poor the waterfall had in effect gone missing.

Even farther to the west lay shadowy Ossokmanuan Lake, visited only by an early surveyor coming from the Quebec side. It might or might not prove to be the source of Unknown River. The lake was known also to the Montagnais, but for reasons they never explained to white men, those natives scrupulously avoided the whole region, calling it simply "bad country."

The whole journey was nearly sabotaged by Gino's adopting a new cornmeal diet for the dogs, as recommended by several trappers in North West River. On the trail, the diet proved disastrous, sickening the huskies. One useless dog that collapsed had to be shot and fed to the others, and Scott, out of pity, released another, hoping it would find its way back to the Hamilton. But on March 6, guided by Gino rather than Michelin, for whom this was unknown country, the men stood beside the massive cataract of Unknown Falls. Stunned by its magnificence, Gino wrote in his diary,

The river plunged over a drop of several hundred feet into a deep, steep-sided gorge. Everywhere were tangled masses of ice and great ice-cliffs, while down at the bottom of the gorge was the seething river.

Michelin and Scott were ready to go home. But Gino insisted on driving farther westward. A kind of manic will came over him. For another week he drove his partners on, as he tried to solve the topography of the headwaters. Throughout the first two journeys, it had always been Michelin who took the lead. Now it was Gino, with an impetuosity that at first daunted Scott:

> He led the way through rough country and smooth without ever looking back to see if we were following him. At first this seemed callous; the action of a man who did not care what happened to his companions. But very soon we recognized it as a high compliment and felt that he took it for granted that we could do as much as he. As usual, he gave no orders, but talked of what we had to do as if it were an order to himself.

In his diary, Gino vowed, "I don't care if we have to half starve and kill every dog we have got [for food]. I am not going back to our advance base on Grand [Hamilton] River until we have finished the work we came to do."

For the first time, bad feelings erupted between the guide and his so-called clients, to which Gino seemed oblivious. In the end he drove his partners farther up the Unknown River, where they discovered three more waterfalls. On March 10 the men came upon five small lakes, which Gino mapped and prosaically named Lakes A through E. They never reached Ossokmanuan, but from a promontory west of Lake E they gazed on its vast surface, and Gino concluded that Ossokmanuan was indeed the ultimate source of Unknown River.

At last Gino relented, though he had still failed to reach the height of land. In four brutal days, the men bashed their way back to the Hamilton, then sped down the river in ten more, dropping off Michelin at his cabin at the mouth of the Traverspine. Only three of their seven dogs were still alive. Upon their arrival at North West River, even Gino, lame with a badly inflamed Achilles tendon, admitted that he had had enough.

*   *   *

Rather than wait for the mail boat, not due until June, Watkins and Scott decided to sledge down Lake Melville to the coast, then south along the shore until they could hail the first ship forcing its way through the ice and try to cadge a ride. That dispirited trek took another month, and was not without its perils.

Yet on that jaunt, Gino had started to muse in great detail about his next expedition. Near the end of his exhausting nine months in Labrador, his head and heart were full of Greenland. Everything he had learned and mastered on the Kenamu, the Hamilton, and the Naskapi, he would pour back into the greatest exploit of his life. Scott listened in weary astonishment.

FOUR

Gino at Home

SCOTT AND GINO returned to England in time to enjoy the pleasures and distractions of a long summer. Wrote Scott (speaking for both): "Now it was grand to be back, incredibly grand." Gino threw himself into family and into "lunches, dances, cinemas and plays." He took Pam to the Oxford balls, which usually lasted from 7:00 p.m. to 5:00 a.m., with limitless food and wine. He recruited Pam and Tony for a short holiday at Margate on the sea, and on the hottest days, for punting and bathing on the Thames. Exhausted from the severity of Labrador, "He made up for it," Scott reported, "by being childishly helpless and particularly gay."

Nanny Dennis was in permanent residence at Onslow Crescent, and Gino availed himself of her services as he had as a teenager: Ovaltine by his bed in the evening, tea in bed for breakfast, and "his clothes brushed and laid out for him." If he misplaced some item, rather than look for itself, he called out like a spoiled child for Nanny Dennis to find it.

At the balls and dances Gino met women who were attracted to him, and though bringing them home to Onslow Crescent would have been awkward under the stern gaze of Nanny Dennis, he evidently indulged in more than one fling. His best friend at Lancing, the relatively ascetic Quintin Riley, went so far as to accuse Gino of fickleness.

He responded, "Good Lord, you can't expect me to like the same woman for more than three months."

"What about if you got engaged?" Riley persisted.

"Oh, I had forgotten that."

The question of Gino's sexuality, though peripheral to his exploratory achievements, cannot be entirely ignored. The way he habitually dressed, in three-piece suits with plus fours, carrying his rolled-up umbrella with him even on sunny days, hints at a foppish, fashion conscious socialite. One cannot imagine Amundsen or Shackleton dressing that way. The scene in Scott's biography recounting his first meeting with Gino in his Cambridge suite, with the other youth clad only in an open dressing gown and a wet bath towel, is provocative enough, especially as recounted by the thoroughly "straight" Scott.

Two stray remarks dredged up without attribution, in books written by Jamie Scott's son Jeremy (who is sometimes gratuitously hostile to his father), and by August Courtauld's nephew Simon, push the suggestion of bisexuality further. According to the former, a Cambridge classmate of Gino's thought him "a pansy." According to the latter, a woman who knew Gino (and who may have dated him) called him "a bit of a ponce."

If Watkins harbored ambiguous sexual feelings toward both men and women, those impulses and predilections may partly explain the privacy he wore as a shield, leaving even Jamie Scott perplexed at how he gained only a shallow grasp of his best friend's inner life, even after nine months of sharing every meal and campsite with him. On an expedition such as the BAARE, "normal" heterosexuality was the default setting.

At the same time as he was dancing through the London night, Gino spent many hours at the RGS, working up his astronomical notes and survey records to produce the comprehensive map that was the principal purpose of the Labrador expedition. Scott was by his side every day, jogging Gino's memory and keeping him focused on the job at hand. In theory Gino had also begun the fall term at Cambridge, but college would never be among his first priorities, and in the end, he never graduated.

In the midst of these conflicting preoccupations, Gino was going through a life crisis of sorts. With his father's inheritance virtually gone, and as the head of the family of three ever since the Colonel had absconded to Switzerland, Gino felt that it was time to figure out what he would do for a living. He actually contemplated the conventional path of becoming a businessman, but, as Scott glossed the prospect, "The idea of going in at the bottom of some well-established business and working his way up gradually year by year did not appeal to him at all."

In the eyes of some of his friends, Gino might well have seemed a dilettante. He had never been a serious student; he had failed his entrance exams for the navy; and the two expeditions he now had under his belt had been glorious adventures, but hardly the kind of thing one could turn into a career. Even such stalwart explorers as Scott and Shackleton had anchored their lives in naval careers, the former in the Royal Navy, the latter in the merchant marine.

Gino was scheduled to deliver his talk about Labrador to the RGS on November 4. For weeks, he struggled to compose his address, which would be published in the *Geographical Journal* as the official record of the expedition. As an alternative to a despised career in business, Gino toyed with the idea of becoming a writer. But while so many other skills came easily to him, authorship did not. To Scott he confided that "he could express his thoughts easily enough until he took up a pen, and then his mind went blank."

Gino's frivolity could seem extreme, even to his friends. The passion for jazz and dancing took on manic proportions. Scott captured the paradox during the London summer season: "Very often he went to dances, perhaps two or three in a single night. In the small hours of the morning he ran home—this running was his exercise—and four or five hours later he would be sitting at his room at the Royal Geographical Society, spruce and wide awake, his mind entirely occupied with the work in hand." Still, as Scott testified, "A stranger at a dance would equally have thought that he lived only for dancing."

As the Boswell to Watkins's Dr. Johnson, during these months in London Scott pondered an even more fundamental paradox. "I was surprised when people told me that I must have got to know [Gino] wonderfully well in Labrador," Scott wrote. "To a limited extent I had. . . . The more we did together the more firmly we were connected and the less likely we were to be surprised by a cause of difference. But we did not even grow sufficiently demonstrative to call each other by our Christian names." (All the more surprising, in view of the fact that on the BAARE, while most of the members addressed each other by their last names, Watkins was "Gino" to everybody.) "He had the habit of keeping, as it were," Scott continued, "his best cards in reserve and of producing those which were only just good enough to win each trick. . . . I knew him in Labrador when playing a particular game, but in England he was more or less a stranger."

At last the RGS meeting came around. The lecture Gino had finally squeezed out of his writer's block turned out to be a smooth, well-crafted account of the nine-month expedition that he, Scott, and (during the summer of 1928) Leslie had completed. While he downplayed the suffering and the close calls, he hyped up the commercial possibilities of Labrador—in particular, the hydroelectric potential not only of Grand Falls but of the twin pairs of falls he had discovered on Unknown River.

This time the acclamation from the audience at the RGS was more muted than it had been two years before, when the twenty-year-old wunderkind had been instantly elected as the youngest member in the history of the society after his presentation on Edgeøya. The president, Sir Charles Close, closed the meeting with a polite tip of the hat: "We congratulate Mr. Watkins, Mr. Scott and Mr. Leslie on what they have done, and we thank Mr. Watkins for his lecture and thank him and his companions for the explorations that they have carried out in Labrador." Gino considered the evening a success. "After the whole thing was over we went off and danced," he wrote to his grandmother.

In his naïveté, Gino failed to recognize that the massive Greenland expedition that was already well-formed in his head required funding that was far beyond the coffers of the RGS. Indeed, until late in the planning stages, Gino paid far too little attention to the problem of raising money to finance it.

By December 1929, only seven months loomed before the start of the BAARE extravaganza. Not only had Gino yet to secure funding, but he had neither chosen his team nor leased a ship to carry men, gear, and food to Greenland in July 1930. The spring term at Cambridge would clearly be a washout, but instead of buckling down to organize what historians would later hail as the most ambitious British Arctic endeavor of the last half century, Gino succumbed to the temptations of his sybaritic alter ego, organizing a skiing holiday in Davos for his siblings and a couple of friends, one of them Scott.

Back at Cambridge in January, Gino signed up for geography as his special subject (equivalent to a "major" in American colleges)—reputed to be the easiest course the university offered, though of course it was the logical choice for an explorer. He also signed up for personal tutelage by Raymond Priestley, whose lecture had awakened Gino to the polar regions in the first place. Priestley professed chagrin at taking on a student with so much exploring already on his résumé, but he also wryly noted that, thanks to expeditions and World War I, it had taken him seventeen years to earn his own bachelor's degree at Cambridge. He wondered aloud whether Watkins might complete his undergraduate education in a shorter span.

By that January, Gino needed only two terms and a final exam to finish. But he barely met the attendance requirements to get credit for the spring term, distracted as he was by the preparations for the BAARE and by the time he had to spend in London working with the RGS and fiddling with logistics.

It was the lure of Greenland that decided the career Gino would pursue as long as he could. As Scott wrote, "Quite suddenly he had found himself as an explorer. It was love at first sight." Against the skeptics who wondered if exploring was "a waste of time if he went on with it too long," he flung back his gut-level rejoinder: "No new country had been considered valuable until it was explored and then exploited. . . . It was something which had to be done." With eerie prescience, he also told Scott more than once "that one must have done everything by the age of twenty-five."

On fire with his new resolve, he quickly drafted a full proposal for the BAARE to present to the RGS. In thirteen paragraphs of urgent exhortation, he put the development of the air-route shortcut from England and Europe to the Pacific Coast of North America at the forefront. Secondary in importance, but a prerequisite for any successful flight, was the establishment of a year-round meteorological ice station at least 140 miles inland at a high altitude on the ice cap. Nothing of the kind had ever before been attempted. These demands in turn would dictate a base camp on the difficult and little-explored east coast of the great island, and a full year in the field. To fulfill all the scientific goals of his exploratory pipe dream, and to ensure the manpower to carry out six other missions besides the Ice Cap Station, he proposed a team of fourteen members.

No one would ever accuse Gino Watkins of modest pragmatism when it came to dreaming up his exploits. It was only near the end of his confident proposal that he dared calculate a budget for the extravaganza. Cutting the costs to a minimum, he could not reduce the total to less than £12,000—an unheard-of $890,000 in today's dollars. The response of the RGS is not on record, but one can imagine the guffaws of the senior fellows of the society.

Gino could hardly have chosen a worse time to propose such a grandiose and expensive expedition, for the stock market crash of October 1929 was sending shock waves across Great Britain and Europe as well as the United States. A more practical man would have drastically reduced the scope of his Greenland plans, or even shelved the whole endeavor, but Gino plunged ahead, as he started to select his team and negotiate for a ship without any looming hope of a way to pay for the expedition. At some point that spring a Hollywood movie company offered to help foot the bill, in turn for a say in how the story would unfold. Desperate, Gino cozied up to the filmmakers until he learned that the script called for "a strong love interest" and the dramatic deaths of two team members by falling into a crevasse.

By mid-spring he had started to assemble his fourteen-man team. First on board was Jamie Scott, who, no matter how worn out and eager to get home he had been as he and Gino had sledged down the Labrador coast, was galvanized by his partner's heady talk of a truly ambitious Greenland program for the next year. The next recruit was August Courtauld. Three years older than Gino, Courtauld had graduated from Cambridge in 1926 and was working as a stockbroker in London. Although they had overlapped for one year at Cambridge, somehow they failed to meet, despite both becoming acolytes of the Cambridge tutor James Wordie. Courtauld would be the only member of the BAARE besides Watkins with prior Arctic experience, having joined Wordie's probes of east Greenland in 1926 and 1929. (Scott would make a third, if one counts Labrador as truly Arctic.) Between his Greenland forays, Courtauld had also managed to fit in a three-man exploration by camel of a little-known sector of the Sahara in 1927.

As soon as Gino and Courtauld met in September 1929, the older man—bored silly by the London Stock Exchange—was keen to join the nascent BAARE. Gino knew that Courtauld was the scion of one of the richest families in England, and August himself, thinking that his father, Samuel Courtauld, might become a major donor to the expensive Greenland plan, invited Gino to a shooting party at the family estate.

The way Gino always told the story of that weekend rings true to his improvisatory bent. He showed up in the middle of the shooting party, whereupon August offered to lend him a gun. But Gino held up his right hand covered with bandages and lamented that he had badly sprained his thumb. Actually there was nothing wrong with the thumb: the problem, Gino later confessed, was that he couldn't afford to tip the gamekeeper.

Still, the weekend seemed a social success. But Samuel Courtauld was damned if he would contribute a dime to the cockamamie Greenland scheme. His son had already wasted enough of his youth on fruitless adventuring: it was time he buckled down to the business of finance.

Already at odds with his parents, August vented his disgust to his fiancée, Mollie Montgomery. "He wants me safe and stuck rather than run the risk that his family might ever get out of the rut of complacent money-making," he wrote her. "I am so sick with my family I can't even sit with it. Let it fug in its own noxious library."

Out of this impasse, however, rose an angel. Unwilling to accept defeat, August signed on as treasurer for the expedition. "I tackled the cousins and aunts," he later reported. "They stumped up nobly." But the angel was Samuel's cousin, Stephen Courtauld, an even richer patriarch and, as the founder of the prestigious Courtauld Institute, a man who cared more about art than he did about business. In the end, Stephen Courtauld kicked in the bulk of the funds that allowed the BAARE to become a reality.

As the BAARE gathered momentum, despite the Depression, no shortage of other applicants offered their services. Gino passed over several experienced men and ended up with a team ten of whose fourteen members had Cambridge connections but little exploratory experience. Five men were "lent" by branches of the army, the air force, and the navy. Only one man among the fourteen was over thirty, while their average age was twenty-five. Among them all, at twenty-three, Gino was the youngest.

He later explained the rationale behind his choices in a statement that to many polar veterans might have sounded specious—though it hints at a mania for control that, as always with Gino, seemed at loggerheads with his casual, laissez-faire style of leadership in the field:

> I have always deliberately chosen amateurs for such expeditions rather than men who have had Arctic experience on expeditions other than my own. . . . I prefer that all members of my expeditions should have gained their knowledge with me, since in that case I always know the exact amount of experience possessed by each member of any sledging party. If anything goes wrong with any one of these parties and it fails to turn up at the proper time, I can judge more easily what the leader of the party will do in an emergency.

Put so baldly, Gino's manifesto sounds autocratic, even a bit pompous. Yet on the BAARE, something very like that calculus of risk would work out again and again.

The last two months of preparations for the BAARE caught Gino up in a whirlwind of logistics. He would dash off to the R.A.F. Reserve aerodrome outside London to brush up on his pilot skills, for in his proposal he had promised to make the culminating flight, solo, all the way from east Greenland to Winnipeg. Meanwhile he negotiated for the purchase of two de Havilland DH60G Gypsy Moth airplanes, at the same time as he hired agents in Norway to find him a suitable ship to lease. In the end he snagged the *Quest*, a sturdy warhorse of a vessel, which had been Shackleton's ship on his very last assault on Antarctica— a journey aborted early after the Boss died of a heart attack on board before the team even reached the southern continent. He was only forty-seven years old.

In late spring Gino learned that he must travel to Copenhagen to secure permission for the BAARE, for Greenland was then a colony of Denmark. (In 1953 Greenland would be redefined as a "district" of

Denmark.) Gino had naïvely assumed that at Angmagssalik, the main settlement on the east coast, he could buy or rent sledge dogs for the numerous jaunts he planned to make onto and across the ice cap. But the Danes told him that the east-coast dogs were small and unused to any toil more strenuous than local travel. Instead, he would have to recruit a member to head for the west coast well in advance of the main expedition, secure stronger dogs there, then transport them to the Faroe Islands, where a Norwegian Arctic veteran maintained a whaling sta tion. There the dogs would be picked up by the *Quest* on its way to Greenland.

The way Gino conscripted Jamie Scott for this thankless job was characteristically seductive. "I wonder if we could find a man who knows as much about dogs as you do?" he mused. Scott: "Accepting the compliment, I could scarcely refuse the job."

All the while, in the midst of buying supplies and sewing bags, gloves, and booties, Gino was questioning the received wisdom about rations passed down by the great polar explorers of the previous generation. Scott's men had budgeted 5,000 calories per man per day for the thrust to the South Pole. That was too few, Gino concluded, as he devised a menu of 6,000 calories per day. Consulting a diet expert at the Lister Institute, Gino determined to allot a full third of each day's provisions to pure fat, since he had learned in Labrador how after weeks in the cold and on the trail one came to crave margarine or lard. While still in London, he conned Scott into trying the new diet out for one week as an experiment.

It was a hard trial. "All the time he was doing his regular office work," Scott remembered, "nearly all the time he felt sick, but he kept his ration down and everyone was satisfied. . . . He swam at midday, ran in the evening and felt foolish when he had to go out to dinner with a little paper bag of greasy food." Months later, on the ice cap, he and Scott laughed, "remembering that week of horror," as they lay in their tent, well-fed and contented.

On July 4, 1930, the *Quest* sailed up the Thames and docked, ready for loading. The day before, at Gino's urging, his sister Pam hosted a cocktail party on board. Now chairman of the expedition committee, Stephen Courtauld arranged a dance to follow the party, attended by all kinds of dignitaries. For the first time, all fourteen members of the expedition (except Scott) met one another. The Norwegian sailors who had brought the ship to London got drunk and climbed the mast.

Two days later, the *Quest* started back down the Thames. For another year, Gino would be separated from Pam, Toby, and Nanny Dennis. "[He] hated these goodbyes," Scott noted. "He tried to dull the thought of parting by talking only of coming home, as it might be next week." But only once the well-wishers were out of sight "could he relax and throw himself whole-heartedly into the life on board the *Quest*. There was no sentiment there."

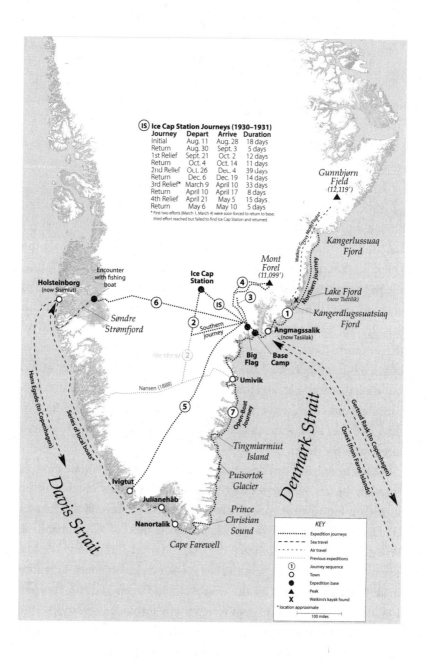

Ice Cap Station Journeys (1930–1931)

Journey	Depart	Arrive	Duration
Initial	Aug. 11	Aug. 28	18 days
Return	Aug. 30	Sept. 3	5 days
1st Relief	Sept. 21	Oct. 2	12 days
Return	Oct. 4	Oct. 14	11 days
2nd Relief	Oct. 26	Dec. 4	39 days
Return	Dec. 6	Dec. 19	14 days
3rd Relief*	March 9	April 10	33 days
Return	April 10	April 17	8 days
4th Relief	April 21	May 5	15 days
Return	May 6	May 10	5 days

*First two efforts (March 1, March 4) were soon forced to return to base; third effort reached but failed to find Ice Cap Station and returned.

Gunnbjørn Fjeld (12,119')

Kangerlussuaq Fjord

Watkins Gipsy Moth flight

Northern journey

Mont Forel (11,099')

Lake Fjord (now Tuttilik)

X Watkins's kayak found

Kangerdlugssuatsiaq Fjord

Holsteinborg (now Sisimiut)

Encounter with fishing boat

Ice Cap Station

Søndre Strømfjord

Southern journey

Angmagssalik (now Tasiilak)

Big Flag

Base Camp

Abandoned (2)

Nansen (1888)

Umivik

Open-Boat Journey

Hans Egede (to Copenhagen)

Series of local boats

Tingmiarmiut Island

Puisortok Glacier

Denmark Strait

Gertrud Rask (to Copenhagen)

Quest (from Faroe Islands)

Ivigtut

Julianehåb

Prince Christian Sound

Nanortalik

Cape Farewell

Davis Strait

KEY

··········	Expedition journeys
··········	Sea travel
- - - -	Air travel
....	Previous expeditions
①	Journey sequence
○	Town
●	Expedition base
▲	Peak
X	Watkins's kayak found
*	location approximate

100 miles

FIVE

Base Camp

No matter how glorious her departure from the London dock, the *Quest* was a seriously overburdened ship. There were not enough bunks for the thirteen expedition teammates and the ten-man Norwegian crew, so several men had to sleep on couches and cushions. Captain Schjelderup had managed to pack an outlandish amount of cargo on the cramped deck, including a crated-up Gypsy Moth, two motor boats, a dozen sledges, all the timber with which the carpenter would build a base camp hut, and tons of gear and food. At Blyth on the Northumberland coast, and again in Iceland, the ship took on a total of fifteen tons of coal, which had to be dumped loose on top of the cargo already on the deck.

The *Quest* had been designed with a flat keel to facilitate passage through the ice, but that meant that even in moderately rough seas she leaned ominously from one side to the other. Once beyond the Thames estuary, Gino promptly got seasick, but so did most of his companions. Five days after leaving London, the *Quest* arrived at the Faroe Islands, where Jamie Scott had spent the last month trying to control the fifty huskies he had bought on the west coast of Greenland. Because the islanders raised livestock, dogs were strictly forbidden on land, so the captain of the whaling station had offered Scott a barge on which to quarantine the ravenous canines as they waited for the *Quest* to arrive.

One dog managed to jump off the barge, reach land, and quickly kill fourteen sheep before he was driven off a cliff to his death.

The twenty-four-year-old Scot Martin Lindsay, a graduate of Sandhurst rather than Cambridge, designated as the team's surveyor, left a vivid description of the greeting the ship got from forty-nine huskies howling in unison:

> The noise became so loud that we were unable to hear each other speak; then, of a sudden, every dog stopped. There was no straggling, no "one cheer more"; the music came to an end to the stroke of time like that of an orchestra obeying the conductor's baton. . . .
>
> It is impossible to tell if it is merely a pastime or whether there is in it some deeper significance—an expression of gratitude, perhaps, or even a prayer. . . . And how they all drill together is as inexplicable as the wheeling of a flight of swallows, or any other miracle of nature.

Getting the dogs onto the *Quest* was a tricky business, as they were loaded onto boats and ferried to the ship. All forty-nine were immediately confined cheek by jowl inside a wire enclosure on the last remaining bit of foredeck. The team also loaded up more than a ton of whale meat from a nearby kill, as food for both men and dogs. With no more room on deck, the meat was hung from the rigging, where it dripped blood on the men and stank abominably. But with the rolling of the ship, all the whale meat fell overboard even before the *Quest* reached Iceland.

Even this jam-packed vessel could not carry all the gear, food, and men for a year in Greenland, so Gino had had to hire a second ship, the *Gustav Holm*, to arrive sometime after the *Quest*, carrying the second Gypsy Moth in a crate and all kinds of other necessities. In any given year, simply reaching the east coast through the shifting and unreliable ice pack was an uncertain business. An expedition led by James Wordie failed to reach land at all in 1923, and his 1929 expedition, with Courtauld on board, took a month to get through the pack, nearly getting frozen in more than once, leaving only three weeks on land to carry out a slapdash skeleton of the program Wordie had planned.

Even before reaching the Faroes, Captain Schjelderup heard grim news over the ship's radio: a Norwegian sealer had sunk in the ice off the east coast of Greenland, taking the whole crew to their graves, while two other ships, damaged by the pack, had given up and limped back to Iceland.

Nonetheless, for the fourteen members of the BAARE, except for the cramped sleeping quarters and recurring bouts of seasickness, life aboard the *Quest* was laid-back. Scott characterizes the camaraderie that slowly built as the men got to know one another:

> The object of conversation was to amuse, to shock if possible and to pass the time between meals. All adjectives were exaggerated, all stories were outrageous and, above all, nothing was serious. There were no rules of conduct. The result, intentional or not, was that everyone became acquainted to the extent that they felt completely at ease in each other's company and knew nothing at all about their private lives. . . . We were not shipmates; we were chaps or something of that sort. One cannot write exactly what we were.

On July 21, fifteen days out of London, the *Quest* reached the outer edge of the ice pack, still fifty miles from the coast. Captain Schjelderup now performed a virtuoso feat of navigating, as from a perch high in the crow's nest he scouted for leads, shouting orders down to the crew steering the ship. The smaller bergs and floes he tried to ram aside, though the shocks vibrating through the ship unsettled men and dogs alike. But when he backed off from an immovable chunk of ice, he also had to make sure not to damage the propeller on the subsurface shoulders of icebergs spreading all around the ship. Gino spent many hours beside the captain up high, determined to learn one more craft of Arctic expertise, whether or not he would ever need to deploy it on his own.

All about the ship, seals basked in the sun on top of floes. They were easy to shoot from the deck, but a bullet had to strike straight into the animal's brain, for otherwise the wounded seal would slip into the water and sink. Using an antiquated Mauser, the best Norwegian marksman slew one unsuspecting seal after another. After each kill, men dashed onto the floes to skin and quarter the prey on the spot. According to Freddy Chapman, who would later write the expedition book, "This seal-meat was more palatable than the whale-meat which we had been eating since leaving the Faroe Islands: both are very tender, but the whale-meat has always rather a strong oily flavour, while seal-meat is delicious." If he spoke for all his BAARE teammates, that was a blessing, for seal would become the staple of the men's diet through the next twelve months.

On the third day, July 23, the *Quest* reached the coast in remarkably good time. "The captain," Chapman recounted, "went off to try and find some Eskimos who could pilot him safely through the maze of ice-filled straits and rocky islands to Angmagssalik." By the next morning, the ship was surrounded by Inuit men and women, overjoyed by the advent of a second ship to relieve the tedium of waiting for the single resupply mission each year that the Danish government allotted to supply the needs of the inaccessible village. Hunters in their kayaks paddled furiously alongside the *Quest*, vigorously throwing their harpoons at nothing, then gathering them up as if to show off their prowess. The larger, stabler, slower skin boats called umiaks carried dozens of women, gussied up with their "amazing top-knots and bizarre clothing."

Angmagssalik, perched on a rocky arm of cove, revealed some dozen wooden houses topped with tin roofs, surrounded by Inuit turf-and-stone summer houses. The largest of only two villages along the immense coast, it served as the administrative quarters of the whole eastern half of Greenland. Some seven hundred Inuit lived in scattered clusters along fifty miles of coast north and south of the village. In residence were only three Danish officials: a storekeeper "with magisterial powers," who also served as governor; a wireless operator with his Swedish wife; and a half-native missionary who was the only person fully versed in both Danish and Inuit.

The first thing Gino did on landing, of course, was to put his gramophone on a level shelf of land and crank up the jazz and show-tune records he had carried from England. A spontaneous dance broke out, the team members lapsing into foxtrot, Lindy, and even the risqué Charleston, to which the Inuit women responded by "danc[ing] a furious double time . . . which was most exhausting."

Although none of the three British accounts (Scott, Chapman, and Lindsay) of this initial meeting breathes a word about it, the moment the BAARE men came on shore they were subjected to a procedure that must have seemed humiliating. Already in Copenhagen Gino had had to pay in advance for an indemnification against any of the team members getting an Inuit woman pregnant: the money set aside would cover the education of the mixed-race child. And now, on the spot, the men had to drop their trousers so that the governor could inspect their genitals for signs of sexually transmitted diseases. All this caution was fueled by an admirable (if in the long run doomed) determination to keep foreign visitors from corrupting Inuit culture. For the same reason, alcohol was banned (though not tobacco).

Gino had hoped to recruit a native family to help the team scout for a suitable locale for the expedition's base camp and to take over some of the household chores. But the governor said that was impossible: "He added that he had lived with these people for thirty-five years, and during all that time he had never succeeded in getting them to do anything for him. He explained that this was not due to any ill will on their part, but that their requirements are so simple that money is no inducement to them to work."

Among the BAARE team, only Watkins, Scott, and Courtauld had any prior experience with the Inuit. The first reactions of the other men typically seized on superficial details, while unconsciously importing the assumptions of the day about racial superiority. Thus Chapman:

> Of the Eskimos themselves our first impression was of men of small stature with copper-coloured faces, dark lank hair, high cheek-bones and amazingly dirty teeth. They seemed to be extremely friendly, and inordinately fond of tobacco. It rather took my breath away to see them knock the hot ashes out of their pipes into their mouths, and then chew them with evident gusto!

Or Lindsay a few days later, as the men started to unload the *Quest*:

> During all this time we had periodical visits from the Eskimos who were living at a small settlement at the mouth of the fjord. . . . We were in those days far too busy to take notice of them, and only found them a nuisance, as they stood about staring with oafish curiosity at all that they saw and fingering everything that they could lay their hands on. It was not for another two months that we had any leisure to appreciate them and realise what delightful people they are.

After two days of scouting, Gino found a fjord thirty miles west of Angmagssalik that he thought should serve well for the team's base camp during the coming year. The place had to satisfy four separate criteria: a bay deep enough to get the *Quest* near shore for unloading; a cove reliably ice-free for the Gypsy Moths to use for takeoffs and landings; a location for the hut on solid bedrock (preferably with nearby running freshwater) beside which the tall radio masts the team would erect could guarantee communication with Copenhagen and London; and a glacial tongue giving access to the ice cap that was not too steep or riddled with crevasses to stymie the dog-sledge teams that would venture inland again and again. Among those four requirements, only the last seemed dubious in the chosen fjord, and indeed, the struggle to get men, dogs, and sledges up onto the ice cap would bedevil the team again and again in the coming months.

There followed two weeks of backbreaking nonstop toil. The dogs were rowed to a small island to be confined until the base was ready for them, and once again their unison howling shattered the silence of the fjord. As it turned out, the *Quest* could get no closer to the shore than two hundred yards, so every item of cargo had to be ferried across in the rowboats, then carried on the men's backs up to the platform a hundred feet above sea level that Gino had chosen for the hut. That included all the planks and doors and insulation that the carpenter, putting in fourteen-hour days, directed to the chosen spot and skillfully assembled to fit the predesigned layout.

The worst job of all was hauling the fifteen tons of coal. As Lindsay wryly wrote,

"Having seldom previously carried anything heavier than a message, I was surprised to discover that, although exceedingly unpleasant, it is not impossible to take a load of 120 lb." Gino organized the hauling into back-to-back twelve-hour shifts, and carried his own share of loads. The coal dust covered the men's faces and got into their eyes. "For a while we found it hard to recognize each other under a mask of grime," wrote Lindsay. "At the end of each shift we used to have basin-and-bucket parties at the stern and try and try to get a little cleaner before turning into our bunks."

The carpenter, a member of the Norwegian crew from the *Quest*, brilliantly organized the construction of the hut. Digging postholes to anchor the seventy-foot-tall radio masts was more vexing. Men would shovel several feet into the unyielding ground only to hit big stones and have to start over.

Within days the hut was assembled. The main room, twenty feet by twelve, was big enough to include bunks for all fourteen expedition members. The well-insulated walls comprised double layers of match-board with felt stuffed between. There was even a kind of attic built above the sleeping room. Between the vestibule-doorway and the main room there were a kitchen, a combined radio room and workshop, a small darkroom, and storage space for gear and clothing. The only nuisance was that each member had to pass through the outer compartments to get to the main room and his bunk.

At first everything seemed to go well. The radio masts were raised into place, and to the delight of Percy Lemon, a captain from the Royal Corps of Signals and a veteran of wireless transmission, communication with Copenhagen and London worked perfectly. Lemon was also putting together a portable wireless kit to take up onto the ice cap. Ideally, messages sent back and forth between base camp and the ice station would add a huge factor in terms of safety.

By August 3, the two pilots recruited from the Royal Air Force, Narbrough D'Aeth and W. E. Hampton, had put together the Gypsy Moth that had been hauled in crates aboard the *Quest*. During the next few days, a series of flights gave the team giddy hopes for the all-important airborne component of the BAARE. Mounting photographic equipment on board, D'Aeth took Gino up to make a series of revelatory images of the nearby coastal mountains and the edge of the great ice sheet. Among their discoveries was that Mont Forel, thought to be the highest peak in Greenland and perhaps the Arctic, was lower than several yet unnamed mountains that few humans had ever seen, let alone dreamt of climbing.

This early success, however, was vitiated by two alarming developments. As soon as they had unpacked the plane, the pilots discovered that the ski undercarriages—crucial to the planned landings on the ice cap—had been left behind in England. Exasperated, Gino got on the radio and all but ordered the de Havilland company near London to ship the undercarriages to Reykjavík, where the *Quest*, returning from Greenland, could pick them up, then steam back to the east coast before the pack closed in to deliver the undercarriages to base camp. Amazingly, this desperate fix-it job worked out.

On one of D'Aeth's photographic flights, however, he returned to the base camp fjord to find that the ever-shifting ice had drifted close to shore, leaving no room to land. Somewhat frantically, he searched for an alternative patch of water big enough for a safe landing, but had to settle for a cove three miles from base. Then one of the *Quest*'s dinghies, equipped with an outboard motor, managed to tow the plane home, nudging and bashing aside pieces of floe as it chugged along. The outlook for future flights thus grew a lot cloudier than Gino had hoped.

On August 9, the *Gustav Holm* arrived at last, delivering the second Gypsy Moth and tons more of gear and provisions. The captain was so desperate to get away, though, that the men had little more than twenty-four hours to unload all the cargo and row it ashore. Then once again, they pulled twelve-hour shifts to haul all the stuff up to the depot beside the hut, a mere hundred feet above sea level.

At last the team was ready to launch the BAARE's central mission: to pioneer a route through the glacier tongue spilling toward the coast, then up onto the ice cap for some 140 miles, where they would build a weather station and man it through the winter. On one of the flights Gino had had a close-up look at that glacial tongue. To his dismay, it appeared to be more crevassed and steeper than he had first guessed.

The reality would outdo all the men's worst preconceptions.

*   *   *

The biggest island in the world (Australia and Antarctica being deemed continents) remains hard to grasp in its entirety. Even those old Mercator projections that wildly inflated the size of Arctic lands somehow failed to give Greenland its proper texture and intricacy. Perhaps the image, repeated over and over, of a thin halo of coastline centered by a blank white blob conduced to the simplification of the place's reality.

Consider a few statistics. At 836,300 square miles, Greenland is nine times as big as Great Britain, three and a half times the size of France. It's bigger even than America's self-styled Great Land, Alaska. The size of the ice cap alone is hard to comprehend. It stretches 1,500 miles from north to south (the distance between Maine and Cuba), with an average east–west diameter of 600 miles. That unknown immensity began to loom as a great exploratory challenge in the 1880s, utterly defeating such stalwarts as Robert Peary and Nils Adolf Erik Nordenskiöld (who had already made the first traverse of the Northeast Passage from Sweden to the Bering Strait) before it succumbed to Fridtjof Nansen and his five companions in 1888, on the expedition that first won Nansen his Arctic laurels. But Nansen tackled the ice cap near its southern end, where it was only 280 miles wide, and that daring thrust into the wilderness left all the essential mysteries of the great frozen sheet untouched.

As the Icelandic sagas vividly recount, the European discovery of Greenland was made in AD 982 by Erik the Red, a short-tempered Viking who was exiled for three years after a typical Icelandic blood feud escalated into Erik's murder of two of his neighbors. (The sagas hint at even earlier sightings of the great island by other Icelanders, but they remain hazy speculations.) Heading west into unknown waters, Erik rounded the southern tip of Greenland before sailing up the west coast, where he made landfall. Delighted with his discovery, Erik spent the rest of his three years exploring the coast before he returned to Iceland as an entrepreneur touting fine real estate in the new country that he named, as a promotional pitch, Greenland (Grœnland in Old Norse).

The sales pitch persuaded a large number of poorer Icelanders, devastated by a recent famine, to follow their Pied Piper in hopes of a new life. Erik set out again at the head of an armada in 985, bound for the west coast. The voyage was a disaster: of the twenty-five ships that set out, eleven were lost at sea. The survivors nevertheless started their new colony with a will, establishing two centers, called the Western Settlement and the Eastern Settlement. Population swelled to the thousands, as the colonists imported livestock and attempted agriculture. The Norse colonization lasted for more than 400 years, but sometime in the fifteenth century, it died out to the last man, woman, and child.

The causes of this catastrophe remain one of the great archaeological puzzles in the colonization of the New World. The worsening climate of the Little Ice Age, beginning after 1250, surely made for tougher times in Greenland. It may be that the colonists clung to a mode of existence based on livestock and planting that was no longer sustainable, until famine and starvation wreaked their havoc. But recent studies argue that the Norse tried to shift toward an economy based more on hunting seals and trading walrus ivory to European markets, but died out anyway. The third factor, which represents an even greater archaeological mystery, has to do with the possibility that the colonists warred against Inuit newcomers arriving from the north, and were massacred or starved to death.

Erik the Red reported no "natives" along the west coast when he arrived: as far as he could tell, this was an entirely new, uninhabited land. But he was wrong.

The immensely complex record of Inuit and proto-Inuit settlement of Greenland depends today on subtle but rigorous excavations of ancient and fugitive settlements, and the picture deepens with each successive wave of research. In the popular mind, the Inuit were a single people, the last to cross the Bering land bridge, who spread relentlessly east across the Arctic shores of Alaska and Canada and eventually Greenland, taking centuries to migrate, but retaining a lifeway centrally based on the hunting of seals. Why they stopped with Greenland and pushed no farther east may have to do simply with the 185 miles of open sea between that island and Iceland at the narrowest gap—a greater gap than the Inuit had to leap between any of the lands they traversed from the Bering Strait to Ellesmere Island.

Were these "Inuit" really a single people? Here the picture becomes even more complicated.

The first humans to discover Greenland were nomads whom the scholars are not comfortable calling "proto-Inuit": instead they lurk under the catchall label "pre-Inuit." They arrived sometime around 2500 BC and managed to spread along the west coast and halfway up the far more marginal east coast for an astonishing seventeen centuries. The Saqqaq culture, under which rubric those pioneers are taxonomized, lived in short passage-dwellings made of stones and turf and in skin tents (only the tent rings survive). They did not bury their dead. Driftwood, abundant then (but not today) served as fuel. From unpromising gray slate they crafted scrapers and dart points. With this aboriginal tool kit, they managed somehow to hunt and kill "no less than 45 vertebrate species," ranging from birds and fish (mostly cod and char) all the way up to caribou and polar bears, seals and even whales. Yet they perfected this apparently marginal lifeway without the benefit of the two irreplaceable linchpins of the much later Inuit peoples whom Europeans first encountered—dogs hauling sledges, and kayaks made of sealskin.

So complex are the variant culture groups whose traces archaeologists analyze along the Greenland coasts that five successive and partly overlapping waves of nomads are identified, from Saqqaq through Independence I and Independence II to Early Dorset and Late Dorset before arriving at the Thule people—the "Eskimos" whom European explorers marveled at, misunderstood, and abused.

Yet the mystery remains. During the 450 years that Icelanders settled the west coast, raised cows and sheep and planted root vegetables, fished and hunted seals and walrus, where were those Thule Inuit? For the medieval records give no hint of rivals vying for the land. Yet today's

Inuit preserve legends of their ancestors descending on these interlopers and wiping them out.

For some 270 years after 1450, only silence from Greenland reached Scandinavian ears. Then in 1721, a Norwegian named Hans Egede set out to solve the mystery. Egede was a remarkable man—explorer, archaeologist before his time, and colony-builder (he sailed with his wife, four children, and forty other would-be settlers). But above all he was an ordained theologian, and his primary goal was to find out if the lost Norse colony still existed, and if so whether its survivors had lost their faith.

Thanks to scurvy, the mission came close to giving up and fleeing back to Norway within the first year. But Egede stuck to his vision, and subsequent relief expeditions, thick with Moravian missionaries, arrived in the following years.

Egede found the ruins of the Western Settlement, but no living Norsemen. Instead he found flourishing bands of Thule Inuit. At once the newcomers determined to convert these heathen to the True Church, and so successful were their efforts that a syncretic blend of Christianity with traditional "pagan" beliefs soon took hold, creating existential paradoxes that Gino Watkins's team would still struggle to comprehend in 1930.

Yet even as Egede and his followers transformed the west coast into a new mixed-race populace, crossing Inuit with Danish and Norwegian, to emerge as "Greenlanders," a further mystery lingered. Where was the Eastern Settlement? And what of the east coast itself, so hard to navigate because of the pack ice? Could the last surviving Norsemen from the medieval colony be somehow hanging on there? Did any people at all live along the east coast?

In 1828 a determined Dane, Wilhelm August Graah, set out to answer these questions. He would spend two years in as frustrating a campaign as would ever be waged along the Greenland coast. During the first summer, Graah was unable to round Cape Farewell even to reach the southernmost stretch of the unexplored shore. After wintering over in the last native village west of the cape, he set out again in March 1829. Ahead of his time in embracing Inuit aid and expertise, Graah led a team of three fellow Danes and fifteen west coast Inuit, ten of them women, who rowed the broad sealskin umiaks that were the expedition's only vessels.

Graah's progress was maddeningly fitful, thanks to pack ice, high winds, and snowstorms. He had to sit tight for seventeen days on one small island, twenty-five on another. But on August 8, he reached his farthest north, at 65°15', still barely one-eighth of the way up the massively indented coast. But here at last he met a band of some fifty Tunumiut,* with whom he spent enough time to establish a rudimentary communication.

By now the West Greenlanders were on the verge of mutiny, so Graah, completely committed to his quest, sent back his three Danish colleagues and all but two women among the West Greenlander crew. Then he convinced those loyal paddlers and several local Inuit to push on to the north. Even those east coast natives had little idea if others among their people lived along the dangerous further shores.

In separating his team, Graah had committed to wintering over. Even in a normal season, the hunting was poor along this narrow strip of land, but the winter of 1829–30 was an especially meager one. At first Graah was able to trade for food from the locals, but when they too began to face starvation, they fled en masse in January, leaving the explorer with only a single loyal family as teammates. He was ill through the rest of the winter and barely survived.

But with the return of spring in 1830, instead of immediately beating a desperate retreat to the safe havens west of Cape Farewell, Graah

---

* The Tunumiut people, indigenous Greenlanders from the eastern part of the island, are also known as Tunumiit (a term used for the population in their own dialect), East Coast Inuit, and East Greenlanders.

resolved to make one more assault on the storms and ice floes. Only after failing, on *eighteen* separate thrusts, to better his farthest north, did he toss in his cards. The return saw him pass through a gauntlet of near catastrophes, at one point with his helpers hauling the sole umiak bodily across ice floes while surviving on pieces of whale blubber regurgitated by sharks.

Graah's heroic but ill-starred venture solved the question of whether Inuit lived along the east coast, but left unplumbed the possibility that Viking descendants still hung on somewhere along that inhospitable shore, as well as how far north the Inuit themselves must range. Alas, his bold eighteen-month probe into one of Greenland's greatest mysteries was regarded by his superiors as a failure. Graah's account of the journey, vivid and perceptive, was translated into English and published in 1837, as *Narrative of an Expedition to the East Coast of Greenland*. It is read today, if at all, only by scholars.

Another fifty-three years would pass before any other Danish team tried to improve on Graah's magnificent failure. Meanwhile, starting in the 1850s, vague rumors reached the west coast of a genuine Inuit village farther up the east coast than Graah had been able to penetrate. It even came with a name: Angmagssalik, in reference to a small fish of the smelt family called *angmagssat*—the capelin, which presumably formed the backbone of the local diet. At last, in 1883, a major expedition under a Greenland veteran named Gustav Holm set out to find this village, as well as to settle the still nagging question of whether Vikings from the original colony had made a stronghold there, in the so-called Eastern Settlement. Another vague rumor hinted at the presence somewhere along the coast of an upright stone inscribed with Viking runes.

Like Graah, Holm built his expedition around the umiak. Four of those boats, rowed by women, were complemented by ten kayaks paddled by expert hunters, as well as by three Danish colleagues of Holm's and a West Greenlander to serve as interpreter. The party numbered thirty in all.

Unlike Graah, Holm managed to pass Cape Farewell the first summer, but only just barely, reaching a point some fifty-four miles up the
east coast. Leaving a depot there, he returned to Nanortalik, an established village just west of Greenland's southernmost point, to winter
over. Only in May 1884 was the expedition ready to launch again. It
took a full month to battle the ice and storms and gain the first foothold
again on the east coast.

As the expedition battled the ice and weather, Holm sent back most
of his team to survey more carefully the coast they had already explored.
One small group of Inuit whom Holm met told him there was no chance
of reaching Angmagssalik that summer, but a single native called Ilin-
guaki, "this extraordinarily clever and kind man," not only contra-
dicted the naysayers but effectively guided Holm onward. On August 25
the reduced team passed Graah's farthest north, discovering intact the
cairn the pioneer had built fifty-five years earlier. And on September 1,
they reached Angmagssalik.

Holm wintered over on the same island the natives had appropriated for their village. He collected no fewer than 700 ethnographic
"objects" that he would take back to Denmark. Counting up the scattered population, he arrived at 431 living men, women, and children.
After months in their company, he summarized his inevitably shallow
grasp of their character:

> They are lively, and are endowed with considerable powers of dissim-
> ulation. They are polite and accommodating in their behaviour to one
> another, but at the same time careful not to offend, reserved and sus-
> picious. Deeper feelings, such as love, devotion, or real friendship are
> seldom met with among them.

After nine months living near Angmagssalik, Holm finally headed
south in early June 1885, but not before attaining a new farthest north
of 66°8′ at the head of deeply indented Sermilik Fjord. Back in Copen-

hagen, Holm turned over his "objects" for professional study and wrote a three-volume report about the expedition (never translated into English).

Holm's exploration laid to rest the surmise that the Vikings' Eastern Settlement lay somewhere on the east coast. Not a trace of Viking presence came to light. (It is now known that the Eastern Settlement lay near the southern end of the west coast—it was "Eastern" only with respect to the Western Settlement farther north.)

Yet, astonishingly, Holm completely failed to comprehend what he was seeing during all those winter months at Angmagssalik. Despite coming across many abandoned Inuit houses, some with dead bodies lying desiccated inside, Holm never quite recognized the ordeal the Inuit had just been through.

The years 1882 and 1883 would come to be lastingly known to the Angmagssalik as the Great Hunger. For two years in a row, the catch in seals and land mammals alike dwindled to almost nothing. As their food ran out, the Inuit killed and ate their dogs. When those were gone, they dismantled their umiaks and kayaks and ate the sealskin of which they were made. And when that desperate source ran out, they cannibalized the dead bodies of their neighbors and even their own families.

Near the end of this devastation, the survivors turned to their all-powerful *angakut*, or shamans. Those seers predicted that in the third year of the Great Hunger, the *Kratouna* (white men) would come and kill all the rest of the Inuit.

During those nine months that Holm's small party complacently wintered over near Angmagssalik, the natives lived in a state of barely controlled terror, as they waited for the prophesy of the *angakut* to come to pass.

Reverberations from that dark time still lingered forty-five years later, when Gino Watkins and his thirteen teammates arrived. But those Englishmen would remain as oblivious to the spiritual undercurrent as Holm had been in 1885.

* * *

After two steady weeks of the most grueling labor into August 1930, base camp for the BAARE expedition was all but established. The last job the men had to undertake before starting the trek up onto the ice cap was to get the dogs off the island and into a wire pen that Jamie Scott had devised. This turned out to be a chaotic business.

Having been sequestered on their island for two weeks, the huskies were wary of change. When the men brought fresh seal meat as bait, "many of the wilder dogs," as Chapman reported, "retired to the farthest parts of the island, standing belly-deep in the water growling at us as we waded in after them." It was an even tougher job getting the dogs into the whaleboats to row them to shore. A few jumped out of the boats and started swimming for land, but had to be driven back for fear they would gain terra firma and go wild, perhaps never to be caught again.

Once all the dogs were wrestled inside the wire pen, they had to be harnessed and attached to the sledges before they could be freed—and not just in any permutation, but in "families" of six or seven that recognized a lead dog as master. Before this could happen, though, Scott and Watkins made a preliminary reconnaissance of the glacial tongue that offered the only route up onto the ice cap. What they discovered was that the steepest part of the slope had been worn to bare ice by the winds. Such a jagged surface threatened to cut the huskies' paws enough to cripple them.

Against this setback, the team had brought leather booties for the dogs. But now, it was discovered that someone (Chapman blamed Scott, "never much of a mathematician") had packed only fourteen booties for the fourteen dogs that would make up the first main team, instead of the requisite fifty-six. Exasperated, the men crafted booties out of all kinds of scrap leather.

At last the pioneering thrust through the glacial tongue, breaching the gauntlet the men would name Buggery Bank, was ready to set out.

Yet despite all the tribulations, Gino was so confident of success that he left that mission to a five-man party under Scott. All the rest of the team, except Percy Lemon, who would stay at base to maintain the radio, boarded the *Quest*. Before her departure from Greenland, she would carry the men on the first of Watkins's six secondary sub-expeditions: to explore and survey the coast northward, far beyond Angmagssalik, as far as she could steam in the time left before heading back to Iceland.

On August 12, the *Quest* set off, with one of the Gypsy Moths on board to augment discovery. Gino guessed that he could afford three more weeks for this journey, so dear to his heart, and as the eight men boarded the ship and said goodbye to the sledging party, no one was happier or more sanguine than he.

SIX

"That Cat's on the Roof Again"

GINO'S PLAN FOR what he called the Northern Journey was ambitious enough. The goal was the fjord called Kangerlussuaq, 300 miles north of Angmagssalik, the deepest inlet in the whole southern half of the east coast. The place had acquired a mystique, for while numerous ships (mostly sealers) had tried to enter and explore its thirty-five-mile-deep recesses, festooned with subsidiary fjords, none had succeeded. Most vessels, in fact, had been stopped by the unyielding pack ice as far as 150 miles off the coast. Yet the Inuit had penetrated the fjord in their umiaks and kayaks, hunted it for its abundant game, and even built their winter houses in its nooks. By 1930, though, no European had ever been inside this Arctic haven.

The plan was for the *Quest* to hug the coast all the way north, while the men surveyed it and compiled a map to improve upon a chart drawn by the great Danish explorer Georg Carl Amdrup, who in 1900 had led an intrepid team of four on the only previous survey of the intricate coast. In addition, with Narbrough D'Aeth as pilot, Gino would make a series of flights in the Gypsy Moth to get a good look at the unknown ranges at the edge of the ice cap and to make aerial photographs for future study. At the end of the month, he and D'Aeth would fly all the way back to base camp. Four of the other men, after bidding farewell to the *Quest*, would be charged with making their way all 300 miles back to Angmagssalik in the *Quest*'s whaleboat, mapping and exploring as they went with far greater precision than could be obtained from shipboard, and living off depots laid on the outward journey.

The journey didn't quite work out as planned. Only two days out of Angmagssalik, Gino revised his plan. The *Quest* managed to enter a fjord the natives called Kangerdlugssuatsiaq—a diminutive form of the great fjord still 220 miles to the north, because it had the same shape, though on a much smaller scale. Here, and in an even smaller inlet just to the north, D'Aeth was able to take off with Gino for the first aerial reconnaissance and photography trips. The latter bay had a freshwater lake at the head of it, which seemed to offer a failsafe emergency landing strip. In 1932, on his fourth major expedition, Gino would make Lake Fjord, as the team named this inconspicuous but genial refuge, his base.

The team had squandered, however, a week of precious time on this secondary objective. Farther north, the coast was still jammed with pack ice. Impatient to return to Reykjavík, and increasingly fed up with the men's pokey process of surveying and depot-laying, Captain Schjelderup threw in the towel and headed east out to open sea, then steamed hard north toward Kangerlussuaq. This meant leaving some 160 miles coastline not only unsurveyed but unscouted.

In a single push of twenty-four hours the *Quest* covered the distance. The eight members of the BAARE on board woke on August 23 to find themselves opposite the mouth of Kangerlussuaq. Watkins and Chapman at once climbed a hill to get a look: "It was an exhilarating thought," mused Chapman, "that we were the first white men ever to behold that view."

The next day, Captain Schjelderup's mood took a full U-turn, as, to his surprise and delight, he was able to pilot the first ship ever to enter the legendary fjord. It was immediately evident that the inlet teemed with wildlife: seals above all, but also narwhals. The captain went off in a whaleboat to try to harpoon one of these mammals, whose single tusk was a hunter's prize, but the narwhal dove deep before the boat could approach. On the coastal sands, the men discovered innumerable tracks of foxes, wolves, and polar bears.

But Gino was avid to fly, and in the vast fjord there was no stretch of water long enough for a takeoff by the Gypsy Moth. So the next day, pursuant to Gino's orders, the *Quest* steamed back out of Kangerlussuaq in search a suitable cove. Left behind to survey the fjord in the whaleboat were Chapman, Courtauld, and Alfred ("Steve") Stephenson.

* * *

Of all the members of the BAARE, F. Spencer (Freddy) Chapman was the closest rival to Gino Watkins in energy, nerve, and ambition. Since it was Chapman who ended up writing the expedition book, *Northern Lights*, it is through his eyes and voice far more than Gino's that posterity apprehends the Greenland campaign of 1930–31.

Born three months after Gino, Chapman lost his mother only a few years later, and his father was killed at the Battle of the Somme in World War I. Raised by a clergyman friend and his wife, Chapman grew up as a misfit who loved rambling solo through the countryside, and at an early age he became a passionate bird-watcher—an avocation that would last all his life.

Chapman attended Sedbergh School in Yorkshire, which he "loathed" (his own word) until a sympathetic headmaster excused him from school sports and allowed him instead to hike and climb in the nearby hills. Unlike Watkins with his indifference to Lancing College, Chapman later in life became a devoted alumnus, and through much of his mature life he cherished the semi-secret hope of being named headmaster of Sedbergh.

Chapman went up to Cambridge in 1926, the same year as Gino. He quickly joined the mountaineering club, and while still an undergraduate, spent one summer climbing in the Alps and another in Iceland studying plant and bird life. At Cambridge he and Gino got to know each other, but their friendship was not a close one before the BAARE. Chapman's account of how he was invited to join the Greenland expedition serves as one more testimony to Watkins's powers of persuasion. In his telling, the two men bumped into each other by chance on the ski slopes of Davos in December 1929. The conversation proceeded thus:

"Hullo, Gino. How's Labrador?"

"Hullo, Freddy, how's Iceland? What are you doing here? Come with me to Greenland."

"Right you are. Why?"

Now, in Kangerlussuaq Fjord, Chapman took charge. August Courtauld was his equal in many respects, but that son of wealth and privilege had a penchant for diffidence and aloofness. As he would a year later on the last of Gino's seven mini-expeditions (and perhaps the most dangerous of all), Courtauld was content to let someone else call the shots. (Courtauld's first impression of Chapman, recorded in his diary aboard the *Quest* on the way to Greenland, was "a charming chap but horribly hearty.")

One of the first tasks the threesome performed was to cache no fewer than forty boxes full of provisions in a tight pile covered with heavy stones. This depot was meant to resupply a sledging party that Gino hoped would reach the head of Kangerlussuaq the next spring. Like several other of Gino's pipe dreams, this one would never come to pass.

The high point of the trio's week in the great fjord was their killing of a polar bear, which at first Chapman mistook for an oversized gull swimming in the water. Giddy with their triumph in felling the beast with a single rifle shot, the men had not bargained for the challenge of simply getting the huge carcass out of the water. For fear of jamming up the propeller by towing the bear, they turned off the outboard motor and rowed instead. It took two hours "of the hardest pulling imaginable" to reach the nearest shore. Then the trio found the corpse too heavy to lug onto dry land: instead they stood knee-deep in the icy sea and skinned the bear where it floated.

The excitement had only begun. Only an hour later, as the men scouted for a camp site, another bear appeared in the middle distance. The men didn't need more meat, but they had been too preoccupied during the first kill to take any photographs, so now they set off in the boat after the second animal in hopes of documenting the domain of *Ursus maritimus*. The unreliable motor abruptly died, and as they fiddled with it, they failed to notice that the bear had doubled back toward them. As Chapman reported,

> Suddenly I looked up and saw that in an amazingly short space of time the bear with great agility had clambered onto an ice-floe and had turned snarling ready to attack. I dropped my camera into the bottom of the boat and seizing the rifle which was ready loaded on the bows, took a snap shot at the bear just as he was about to leap into the boat. We were so near that as he fell stone-dead he almost upset the boat with the splash of his huge body.

The low point of the week, on the other hand, came about as a stroke of beginner's carelessness. Having tied the boat to the cliff, the men woke on August 29 to find it full of water with all its cargo missing. Evidently the boat had risen with the tide, but on sinking with the ebb, its gunwale had caught on a point of rock and tipped the boat on its side. "Not only had our petrol supply, ration boxes and a miscellaneous collection of oars and poles gone," Chapman recorded in dismay, "but our instrument box, without which our survey was impossible, had completely vanished." Missing were the sextant, theodolite, field glasses, and several cameras.

For ten hours the men peered into the green water looking for their lost valuables on the floor of the fjord. They recovered much of the debris, some of which still floated, but the all-important theodolite box could not be found. Chapman gave up and started cooking bear steak, chastising himself for his carelessness, when suddenly his two companions cried out that they had spotted the box. But it lay fifteen feet below the surface of the water. The men attached a boat hook to an oar and tried to snag a handle of the box, to no avail. There was a single last resort.

Martin Lindsay, who heard the story later, described the rescue of the box: "Chapman then said he would dive down and open up [the handle]. They were all very doubtful whether he could manage to do so in such cold water. He stripped and dived off a rock, and the others watched the white gleam of his body going down and down until it blurred against the bottom. At last he rose to the surface and said that he had done what was required, and the box was now hooked up without further difficulty." The theodolite survived its watery entombment.

On September 2, Watkins and D'Aeth succeeded in flying the Gypsy Moth all 300 miles back to base camp. As Captain Schjelderup picked up the trio in Kangerlussuaq Fjord, conscience got the better of his impatience. He told the six remaining men that "it was quite suicidal to hope to get back to Base in our small boat." Instead, he would carry the men home in the Quest, even if it meant a further risk of getting trapped in the ice.

During the thirteen days of that return voyage, a daily tension between the captain and the team played out, Stephenson's madcap surveying versus the skipper's full speed ahead. In the end, the map compiled by the BAARE team was marginally better than Amdrup's. But it was a map riddled with holes—lacunae, where the men had had to bow to Captain Schjelderup's benevolent haste and leave miles of coast uncharted.

On September 14, the Quest reached base camp, where a joyous reunion broke out.

The cardinal triumph of the Northern Journey, of course, was the exploration of Kangerlussuaq Fjord. Yet the deepest revelation the inlet afforded was one the men barely noticed. Everywhere along the inner coasts of the fjord, the men had come across abandoned Inuit houses and graves. Seeking the simplest explanation, the team hypothesized that with the rise of Angmagssalik as the principal east coast village in the 1890s, complete with a missionary, a Danish trader, and a yearly visit by resupply ship, the inhabitants of Kangerlussuaq had migrated south to that outpost, seeking a more secure existence.

Subsequent research upends that theory. The discovery of a dateable quartz hand scraper proved that "proto-Inuits" or "pre-Inuits" had lived in the fjord as long ago as 2,000 years, and there are also indications that the fjord could have been occupied around 2000 BC. Other excavations documented continuous habitation from at least the fourteenth century AD through the eighteenth. Any depopulation of Kangerlussuaq, and possible migration south toward Angmagssalik, thus long predated the pioneering voyages to the east coast by Graah in the 1820s and Holm in the 1880s.

As the BAARE team found it in 1930, Kangerlussuaq was still a hunter's paradise. What had driven all those earlier inhabitants away? Where had they gone? Even in 2022, no researcher can come up with a definitive answer.

* * *

Before the *Quest* had headed north along the coast, the eight team members aboard had bid farewell to the five men who would make the pioneering trip up through Buggery Bank and onto the ice cap. It would be their mission to find a site for the ice station and to erect a camp there. As this was the first time since arriving at base camp in July that the whole team would be divided, the parting carried a certain emotional freight. But Martin Lindsay, one of the five and the best chronicler of the first ice cap journey, throws cold water on the melodrama. Gino himself set the tone:

A cinema director would have been bitterly disappointed. Strong men did not grip each other by the hand and look across into moist eyes. Nor did the leader in a choking voice say, "Good-bye, old man; I'm sure you'll pull through." Instead he waved his hand, called out a flippant "Cheerio," then turned on his heel.

The five were Lindsay, Jamie Scott, Quintin Riley, John Rymill, and Edward ("Ted") Bingham. Riley was Gino's old chum from Lancing and Cambridge, his best friend in adolescence. Bingham had earned his medical degree at Trinity College, Dublin, in 1926 and then joined the Royal Navy; he came aboard the BAARE as the team's doctor. Rymill, Australian by birth, came to England to study at the RGS and the Scott Polar Research Institute. In all likelihood, it was Frank Debenham, the veteran of Scott's last expedition and along with Raymond Priestley the co-founder of Scott Polar, who introduced Rymill to Gino. Rymill was a man of many talents: a competent mountaineer as well as a pilot who had trained with the de Havilland company, the makers of the Gypsy Moths. On the five-man mission onto the ice cap, Scott was in charge, while Rymill was effectively number two, in charge of all the details of navigation.

Lindsay half-jokingly illuminates the social split that thus separated the five men during their journey. Scott and Rymill appropriated the "boss tent," spreading their maps and navigation tables around the interior, luxuriating in "comparative comfort." Lindsay, Bingham, and Riley were "tightly packed in the other [tent], in a rare muddle." As they got used to their cramped quarters, they "found that we had far more room than we thought, and that it was not really necessary to put an elbow in the next man's face to settle down in a sleeping-bag."

Of the five, only Scott had any experience with dog sledging. From the moment they left base camp on August 11, the men soldiered through a crash course in that essential art. With his penchant for the mock-heroic, Lindsay insists that the idea that it's difficult to learn how to drive dogs is a myth. Yet the slapstick comedy of his account belies that conceit. Simply harnessing the huskies for the first time proved a challenge:

> You get short of breath and out of temper, and you may well think that you will never get the better of such a disunited team. The dogs scamper off when you want to harness them, and fight as soon as you have got them together again. You run forward to deal with the most offensive, get caught round the ankle by his trace and measure your length in the snow. And when on your feet again, you very likely find that the dog which you were going to punish has bitten through his trace and cannot be caught.

To tackle the bare, jagged ice of Buggery Bank, the leather dog booties were essential. But: "Every boot has to be tied on by its ribbons with the greatest care, and so badly did we do it the first time that before we had gone fifty yards Scott's was the only team that had not shed its footwear."

Scott taught his comrades to use the Inuit commands the dogs had learned since they were pups in West Greenland: *damma* for "pull," *illy* for "right," *yuk* for "left," and *unipok* for "stop." Similarly, the men overcame their reluctance to use the whip, for under its sting the huskies had first learned to haul their sledges. "To get the best pace out of a team," Lindsay declared, "you must be able to use a whip with either hand sufficiently well to lash any part of an offending dog, and there must be an understanding between master and team as profound as that between huntsman and hounds."

After four or five days, the men became fairly competent handlers of their teams. But during those days, they faced the "monster" they had named Buggery Bank. "I for one thought we should never get our sledges up there," Lindsay confessed. The men donned harnesses themselves to help the dogs pull. Sledges routinely overturned, and it took two men to right each one. Scott, wearing his rugby vest from Cambridge, inscribed "Varsity Match of 1928," took charge. After a monumental effort spanning two full days, the men succeeded in wrestling all four sledges and eleven loads to the top of Buggery Bank, at 2,000 feet above sea level.

To make matters worse, in the midst of the struggle, Rymill developed a "poisoned wrist" so painful that he could not help with the hauling. Bingham, the doctor, lanced the wound, while determining that had he delayed another day, Rymill might have died, for the infection had already spread to his shoulder. Now the man was reduced to lying in his sleeping bag below the Bank, listening to his teammates grunt with their burdens and curse the dogs. Once all the sledges had reached the top, Rymill barely managed to hike up the icy gauntlet without assistance.

As men on expeditions tend to do when they're working really hard, the five got along well. Lindsay makes further fun out of the tight quarters in the tent relegated to the second-tier trio:

> When after supper the three of us got into our sleeping-bags, we exactly covered the floor space so that there was nowhere to put the candle, which hitherto had stood precariously on the edge of a ration box. We each took it in turns to hold it while the other two settled themselves. Then the candle was blown out, and we wriggled further down in our bags. A passing thought claimed my attention, but before I could follow it up I was asleep.

One evening, after the men had settled in, a curious dialogue ensued. Lindsay transcribes the speakers as "A" and "B," but they are obviously Riley and himself, respectively.

A: "Bother! I've forgotten my prayer-book."
B: "But dammit, you don't want to say prayers on the ice-cap, do you?"

A: "Of course I do."

B: "And can't you say prayers without a prayer-book?"

A: "Yes. Only I like to remember the saints' days; it makes it so much more interesting."

Born and raised in Cornwall in a strict Anglo-Catholic family, Riley was serious about religious matters even at Lancing and Cambridge, despite being a happy conspirator in Gino's hijinks. Short, slender, and inclined to pessimism, he nonetheless became a stalwart on the BAARE. At base camp, he took charge of the Sunday services. His teammates may have teased him about his piety, but they knew not to cross a certain line. Farther up the ice cap, Lindsay needled his partner again, asking him why he needed a prayer book in the middle of the wilderness of Greenland. "In case you die," was Riley's chilling answer.

Scott later remembered the disparate nature of this odd couple. Lindsay referred to himself as a typical soldier, but this belied a more colorful truth. He adored travel and big-game hunting and spent weekends in England hunting from his country house. He was genial and careless with his personal possessions—hardly the archetype of a soldier. Riley, on the other hand, had been a coxswain on his crew at Cambridge, and obsessed over yachting: the only passion, apart from his family, to rival religion. The meticulous Riley fussed over creature comforts the others did without on the ice cap. He justified the weight of a rubber hot water bottle, though the water was commandeered in the mornings to make porridge by his fuel-conscious companions.

Physically, too, they were a Mutt-and-Jeff mismatch: Riley light and thin enough to serve as a coxswain on a crew; Lindsay the tallest, and with Rymill the strongest member of the BAARE.

Above Buggery Bank, the five men faced a new challenge, vexing in an entirely novel way. The glacial tongue that culminated its plunge in the Bank stretched far to the west, and it unfolded as a maze of hollows, small ice towers or seracs, and crevasses hidden by snow bridges. In the warmth of August, the hollows were filled with slush, and as they made their way through them, the men often waded ankle-deep in watery snow. Each night's drying of boots, socks, and trouser legs over the Primus was undone the next day by another stumble through the slush.

Restored to relative health, Rymill now took the lead, scouting a route through the labyrinth. When the men entered a crevasse zone, three of them roped up while the other two followed, relaying the sledges. "The leader probed in front of him with a long stick," Lindsay wrote, "while the other two were ready to hold him up if he fell through."

At first the men were terrified of the crevasses, which looked "hundreds of feet deep." Except for perhaps Rymill, with the tamer version in the Alps, none of the five had ever had to deal with those perilous fissures rent by a glacier's flow. At first, as Lindsay put it, "When you step over a narrow one you look down into the blackness of that great yawning gap between your feet, and can picture yourself falling to the bottom." But after a day or two of crevasse-skirting, "you realise that it will be entirely your own fault if you do go down one, and your confidence is in a large measure restored."

Just as they thought they had crevasses figured out, however, they ran into more fiendish configurations. Once the men had to backtrack three-quarters of a mile after hitting a dead end of crisscross slots; on another day, both Rymill and Scott got bad scares when each stuck a foot through a snow bridge before lurching backward. On August 17, all seven dogs pulling the lead sledge broke through another bridge and plunged headlong into a crevasse. Remarkably, the harnesses held, and the men were able to haul the howling huskies one by one back to the surface.

All this drama took its emotional toll on the men. But on the same day, at last they escaped the crevasses and sledged onto the ice cap proper, where the going promised to be faster and safer. In six days, they had progressed only fifteen miles from base camp. Already they had consumed more of their rations than they had anticipated. Now they stopped to build the all-important cache that would come to be known as Big Flag Depot. To lighten their loads, they stashed extra boots and any garments they thought they could spare on the forward push. They also shot one dog, "an excitable animal which refused to do any work," and cached its carcass as emergency food.

Ahead of them stretched the great emptiness. Chapman later evoked the solipsistic miasma every team would enter as it started across the ice cap.

No rock or patch of earth nor any living thing broke the monotony of this featureless plain of dead white. The moving tracery of the high cirrus clouds above and the shadows of the snowdrifts below were all that anyone walking ahead of the sledges could focus his eyes upon.

The plan devised by Watkins was to angle northwest until reaching the sixty-seventh parallel, then march farther west until a site some 140 miles out of base camp could be chosen to serve for the year-round weather station. Navigating across that blankness, however, was no easy task. In the lead, Rymill set his course by compass, but since magnetic north here veered 45° west of north, he simply followed the vector along which the needle pointed. Behind him, Scott kept track of the sledge-wheel to measure distance gained, foot by foot. Coming third, one of the others took bearings on Scott and Rymill with his own compass to ensure that the men didn't veer right or left. Every half mile, the team attached a one-foot-square red flag to a four-foot bamboo pole to mark the route both for the return and for future parties. Each flag bore a number, marking a sequence of more than 200 between Buggery Bank and the eventual weather station.

At Big Flag Depot Scott had reduced the dogs to three-quarters' rations. He worried constantly that his team would have to cut into the food boxes intended to sustain the pair of men who would occupy the station for the first stint. Every day the men gained altitude, and they keenly felt the thinning of the air in terms of how much more quickly they got out of breath as they sledged forward. But at last they were making good time, covering fifteen to twenty miles on the best days.

It was here for the first time that Gino's radical revision of the standard polar diet paid its dividends. As Lindsay rhapsodized over the usual dinner,

> The greasy brown lumps of pemmican were put to float in the pot in the company with the greasy yellow lumps of margarine; two spoonfuls of Plasmon "Nerve Tonic" and three of pea-flour were added, and then in ten minutes we had a hot meal that we thoroughly enjoyed. Pemmican is concentrated beef with added fat, and no one could call it a delicacy; margarine, which with the pemmican makes up half the ration, is known to be—well, not butter. But after a hard day's sledging in cold weather, it always seemed to us that there could be nothing quite so good anywhere as those plates of hot stew.

This was the same food that Gino had spent a week testing back in London, carting his brown-bag suppers to fancy dinner parties and fighting down the urge to chuck up the greasy mess in front of fellow guests. By now the virtue of a fat-rich regimen for men (and women) working hard in cold climates is an axiom of polar travel.

On August 20, the monotony of the ice cap was relieved by the shimmering image of a range of mountains far off to the right. Here, the men knew, stood Mont Forel, destined to be the goal of a climbing campaign the next spring. But with the dog food dwindling daily, the men deviated not at all in their course, tempting though it might be to get a closer look at those unexplored peaks.

On August 23, Scott and Rymill calculated that they were only fifty miles short of the predetermined goal of planting the ice station 140 miles in from the coast. Of course no matter how accurate the sledge-meter, measuring distance gained during the zigzag course among the crevasses was bound to introduce errors. The team had also crested 7,000 feet above sea level, and still the plateau rose gently ahead of them. The altitude came as no great surprise, for the men knew that on Nansen's traverse of the ice cap farther south, his team had reached an apogee of 8,900 feet. The question remained, however, whether the ice cap swelled uniformly to such heights across its great expanse, or instead might be punctuated by deep valleys. Old tales circulated among the West Greenlanders hinted at ice-free oases far inland, full of vegetation, animals, and even (outlandish though it seems) tribes of humans never before contacted. Two months hence, Gino himself would lead a two-man jaunt setting off at a right angle from the flagged route to the Ice Cap Station, hoping to intersect Nansen's route more than 200 miles to the south and thereby gain a partial answer to the question of hidden interior valleys.

During the ten days after leaving Big Flag Depot, the team covered, by sledge-meter reckoning, 112 miles. They ought thus to be almost 130 miles inland from base camp. On August 27 Rymill got out the time-signal device, which to the men's delight worked perfectly, giving them an accurate reading of longitude. The verdict corroborated the sledge-meter: they were only ten miles east of the arbitrary site where the weather station would be erected.

Scott decreed that the team would start early the next morning and head straight west before pausing to take another time-set measurement. The altitude read 8,600 feet. But after seven miles of sledging, the team realized they were on a barely perceptible downward slope. For days Scott had worried about dog food, realizing the huskies would have to make the return journey on half-rations. Already the dogs were eating their harness traces as they lay staked out at night, and all the dog pemmican had to be hidden inside the tents to keep the canines from attacking it in their ravenous hunger.

Without much ceremony, Scott declared a halt. Right here the Ice Cap Station would be built. If it stood closer to 130 miles from base camp than 140, so be it: the number had been arbitrary from the start. At 8,200 feet, the station would be high enough to give the two men occupying it an ideal perch from which to measure everything to do with the complex weather patterns that governed not only Greenland itself, but the Arctic climate stretching from Davis Strait to the Norwegian Sea.

Those two men were Lindsay and Riley, the odd couple. As quickly as they could, in temperatures just above zero Fahrenheit, the five men set up the big double-walled tent with the trap door in the floor that Gino had designed. All five men spent a last night together, sharing a farewell supper. Early the next morning, Scott rallied Bingham and Rymill to pack only three of the four sledges with a bare minimum of supplies for the dash back to base. They loaded up their surveying equipment, their personal gear, a single tent, a single ration box for themselves, and rations for the dogs for only four days.

Scott's haste was dictated by a well-earned anxiety. The parting at 4:00 a.m. on August 30 was perfunctory at best. Standing outside the domed tent, Lindsay and Riley watched the sledges dwindle into the distance, "until they disappeared in a gentle undulation in the snow."

* * *

Seldom had two explorers found themselves so suddenly alone, so far from the nearest human refuge, in such a tiny island of purposeful existence surrounded by (to quote Robert Frost) "A blanker whiteness of benighted snow / With no expression, nothing to express." As if to ward off that emptiness, Riley and Lindsay at once put themselves to work. The first job was to enlarge the tunnel that led from the door in the tent floor to its exit beyond the tent. From a cramped initial passage "through [which] we used to scramble with much leverage of the elbows and sometimes a little profanity," they carved a six-step staircase downward that led to a ten-foot corridor at the end of which they dug a new exit straight up. Next the men positioned close to the tent all the gauges they had carried up on the sledges; eventually they built small "snow-houses" to shelter each one. These included a "comb nephoscope" to measure cloud speed, a pair of maximum-minimum thermometers, poles to measure snow depth, a barograph to record barometric fluctuation, a thermograph to record long-term temperature variation, and an anemometer to annotate wind speed (Lindsay called the device a "whirligig"). When the first mild storm piled a snowdrift against one wall of the domed tent, the two men built a five-foot-high snow wall, which ultimately proved useless to forestall drifting.

Curiously, Lindsay and Riley had been given no clear date by which they could expect to be relieved by the next team to climb the ice cap and the second pair to take over the station. Their ration boxes held enough food for five weeks of normal consumption, but perhaps half-wishfully, they began to expect a team to arrive after only three weeks.

After the grim toil of the nineteen-day push to reach the site of the weather station, the relative ease and leisure of the pair's tasks as monitors seemed almost luxurious. As Lindsay later wrote, "This funny little dwelling was a very happy home for us, and Riley and I look back on the days we spent there as being amongst the most enjoyable of the whole expedition." Their only duties were to go outside and read and record the gauges six times a day, at regular intervals. At first the men conscientiously observed that schedule, but they found that they had left their only alarm clock at base camp. It was not long before the 1:00 and 4:00 a.m. observations started to be skipped.

One of the first things the men did was to raise a big Union Jack on a tall pole next to the tent. Though they could never have anticipated the outcome, that single act would prove absolutely crucial more than eight months hence.

Inside the tent, Riley and Lindsay spread reindeer skins that nearly covered the floor, with one week's ration box perched in the middle. On top of the skins they laid out their reindeer-skin sleeping bags, which "had the disadvantage of shedding their hairs into all things" (including breakfast and dinner). Thanks to the men's body heat, and to the compacted snow just beneath the thin canvas floor, moisture seeped into skins and sleeping bags alike, "until every few days they became a sodden mass which had to be dried over the primus stove."

The men had not been so parsimonious of cargo as to exclude a small library of books to fill their idle hours. They had with them Thackeray's *Vanity Fair*, De Quincey's *Confessions of an English Opium Eater*, Sir Walter Scott's *Guy Mannering*, Austen's *Mansfield Park*, the discourses of Socrates, as well as, curiously, Fowler's *Modern English Usage*. From the start, though, Riley had his priorities. On their first full day in the domed tent, he wrote in his diary, "I hung my crucifix over my sleeping-bag and we hoisted the Union Jack outside, and so we have both Christian and National emblems erected." Lindsay responded by reading aloud from *The Oxford Book of Modern Verse*, which he called the vade mecum of the small library.

The two men had counted on an Aladdin lamp that burned paraffin for both heat and light, but they couldn't keep it lit. It would flicker to life briefly, then sputter out within seconds. For a full week, wrote Lindsay, " 'lighting the lamp' was a game that we played every night after supper," until they gave up and counted their candles, finding just enough to stretch through their anticipated occupancy. Still, it was a real loss. Only weeks later did men from the second ice cap party take apart the lamp and discover that a spare burner, wrapped in paper and secreted inside the lamp's stem, was the cause of the problem.

Perhaps the pair's most prized possession was a chess set. They played one or two games each night after dinner. Lindsay rhapsodized the set, "which gave us unlimited pleasure. It made us forget the number of weeks we had slept in our clothes and our few other little trials." It's hard to tell which man was the better player, but one suspects it was

Riley, as hinted at in a September 26 entry in his diary: "Two games of chess tonight both of which I won. M. lost his queen in both of them. He doesn't seem much use without her."

While Lindsay devoured the *Oxford Book of Modern Verse*, Riley turned to his Bible. On Sunday, September 14, "I read a chapter of St. Paul's Epistle to the Romans and St. John's Gospel, also two or three chapters of [Thomas à Kempis's] *The Imitation of Christ*." But once again he regretted having left his prayer book at base camp. Later during his stay, he would turn to *Wuthering Heights* ("a delightful book") and *Jane Eyre* ("hard to put down").

Squatting cross-legged on the floor, Riley did most of the cooking. By now, in their relatively sedentary state, the two were growing sick of pemmican. "It keeps the body twitching," Lindsay insisted, "but does not comfort the soul." Riley had brought along a small supply of dried peas and prunes, which he added to the evening stew every other day, but the variation only slightly relieved the monotony of life inside the domed tent.

The same monotony afflicted the pair's conversation. Later Lindsay could make a joke of it:

> Before long Quintin and I got to the stage of having told each other the same thing several times over. We made a pact that on these occasions the listener was never to interrupt, but to hear the story through to the end and show a fitting appreciation. Every now and again the "Good heavens! Not really!" would sound slightly unconvincing and you would discover that it was the third time that week that your companion had suffered the same rather uninteresting account of something you had once said at your private school. Fortunately neither of us ever told golfing stories.

In *Those Greenland Days*, Lindsay's deft memoir of the BAARE, published in 1932, he sounded the rueful lament of expedition friends who forge a bond under shared privation and achievement that sadly cannot survive the moment: "And although the days we spent together on the ice-cap broke down all barriers so that we learnt just about every-thing about each other, we have, strange to say, never returned to the same intimacy."

September, the men knew, was one of the warmest months of the Greenland year. But as the days grew shorter and the temperature crept lower inside the big tent—too large to heat adequately with the Primus alone—they began to mind the loss of comfort. By September 13 they were recording temperatures at dawn as low as twenty-five below zero Fahrenheit. Lindsay's diary on successive days tersely recorded the new annoyance: "chilly," "cold," and "damnably cold."

By September 27 Riley and Lindsay decided that the relief party must be a week overdue. To deflect their anxiety, they traded quips that made light of their situation. When new snow on the tent roof melted and slid off, they would say, "That cat's on the roof again." Each evening, "The postman's late again," and in the morning, "Is there anything in the paper, dear?"

Yet despite these jokes, the men started to fantasize about disasters that might have incapacitated the relief party. The base camp hut might have burned to the ground. Or the whole relief party, somewhere in the maze above Buggery Bank, might have plunged together into a huge crevasse.

Explorers are used to having some control over their fate. Nothing is harder for them than to be at the mercy of events beyond their knowl-edge or their skill to alter. (Waiting on a glacier without a radio for a pre-arranged airplane pickup on a vaguely specified date is a classic example.) Thus after September 27, every time either man went outside the tent, he voraciously scanned the eastern horizon, hoping to pick up the distant black specks of an approaching party. "When on his return to the tent he was asked if there was any sign of them," Lindsay wrote, "he would reply in the negative in tones of extreme astonishment."

Should the relief party never appear, the two men schemed, they would wait until "we were reduced to the sweepings and scrapings of our last ration box," then try to hike all 130 miles back to base camp. They had a single sledge they could man-haul, and a single small tent, so they reasoned that they would "haul down the flag and walk out without dishonour. . . . [N]either of us intended to be martyrs to meteorology." Though they talked a brave game, they surely knew that without food, such a trek would be impossible.

Riley and Lindsay had been told that the second pair to man the Ice Cap Station would most likely be Percy Lemon and John Rymill. As early as September 23, four days before the men judged that the relief party must be a week overdue, Riley wrote in his diary, "No sign of Lemon and Rymill." On September 29 a four-day snowstorm swept the plateau. With visibility curtailed, the two men didn't bother to gaze eastward for the black specks, and they rationalized that even if the relief party was nearby, it would hunker down rather than try to locate the station in the storm. Still, on October 1, the men "cleared up the yard most of the day," like housewives preparing for guests, and Riley couldn't resist adding in his diary, "No sign of travellers."

The next afternoon, as the men lay in their sleeping bags, they suddenly heard the whimper of a dog. They dashed outside, soon to be greeted by Scott and Watkins, followed by four other teammates: Rymill, Chapman, Bingham, and D'Aeth. The joy Riley and Lindsay felt was overwhelming, though later Lindsay would insist, "We never had the slightest doubt that we would be relieved in due course."

Once the dogs were removed from their harnesses, all eight men crowded into the domed tent. Riley and Lindsay caught up on five weeks' worth of news. They learned that Scott, Rymill, and Bingham had made a brilliant dash back to base camp after parting from the station men on August 30, covering in only four days a route that had taken nineteen to establish, sledging breakneck along the gradual downhill as far as forty-five miles in a single day.

Inside the tent, every man lit up his pipe (except perhaps Gino), and "the air soon got very foul." Still, they could not get enough of one another's company, so "we complained about it and went on smoking pipes until we all dispersed with aching heads." Watkins and Scott slept in an igloo that Riley and Lindsay had built, while the other six shared the suddenly limited floor space inside the domed tent.

The next day, back inside the tent, the men spent hours chatting and singing songs, lit up their pipes again, "and gave ourselves headaches that were, if possible, slightly worse than those of the night before."

It turned out that rather than Lemon and Rymill, Gino had designated Bingham and D'Aeth—the doctor and the Gypsy Moth pilot—to serve the next ice station shift. On October 4 the others left the weather station. Lindsay, Riley, Rymill, and Chapman started back toward base camp. Scott and Watkins, however, headed off south-southwest, at right angles to the flagged route, on Gino's cherished traverse along the spine of the ice cap to determine whether any hidden valleys lurked unknown between the sixty-seventh parallel and the sixty-fourth of Nansen's pioneer crossing in 1888. Before the two parties went their separate ways, mindful of the difficulties the second relief team had faced, Gino instructed Chapman to launch a third group to start up Buggery Bank, through the crevasse fields, and onto the ice cap as soon as those four reached base camp. If Bingham and D'Aeth's stint as monitors turned out to be shorter than Lindsay and Riley's, no harm done.

Slowed by a blizzard that trapped them for three days, the four returnees needed eleven days to cover the distance Scott's party had covered in four a month earlier. Since they had only one week's food with them, both men and dogs went on half-rations for the last eight days.

All the way back, Lindsay and Riley nursed their secret delight that they had served their sentence in September. October was bound to be much colder, the daylight much shorter, for D'Aeth and Bingham. And then—what would it be like to man the station through the true winter months of December, January, and February?

But as they sledged homeward, Lindsay and Riley thought only about what awaited them: reunion with the rest of their teammates, and indoor comfort in what Lindsay called "the comparative fleshpots" of base camp.

SEVEN

Autumn with the Inuit

DURING THE FIRST two months of the expedition, the only team member who was in constant residence at base camp was Percy Lemon, the radio operator. Though he had met Watkins at Cambridge in 1928, the two were not close friends before the BAARE. At thirty-two, Lemon was the oldest man in the team. He remains something of an enigma.

The fjord in which base camp was situated was also home to several Inuit families, whose summer tents were pitched as close as six miles away. Intrigued by their new neighbors, the locals visited the base camp hut day after day. From the start, Gino arranged to pay the hunters for all the seals and sea trout (which they speared in the water) they could catch, but the supply was never equal to the demands of fourteen men and forty-nine dogs.

Taking advantage of the regular visits, Lemon assiduously set himself the task of learning the Inuit language. Lindsay recounts the frustration of his efforts.

Lemon had to start from absolute zero. It took him several days to get a single noun from them. Whenever he pointed at an object and raised his eyebrows, they thought he was crazy, and laughed. And when they had grasped what he was driving at and named all the things he pointed to, it took him a considerable time longer to obtain the first verb.

None of the fourteen BAARE members ever became conversant in the native language, but several made serious efforts, none more diligent than Lemon's. Scott and Watkins had met Inuit speakers in Labrador, but their dialect would have been barely intelligible to the East Greenlanders.

To their credit, several of the BAARE men soon recognized the complexity of the language. As Lindsay cautiously ventured, "It is perhaps an exaggeration to say that every verb changes both its root and stem in every case of every tense, but this is certainly the principle upon which the language is built." And the men were impressed by the sheer size of the Inuit vocabulary. Lindsay cited not the hoary cliché of "fifty words for snow," but rather that "they distinguish between the swimming of man, dog, bear and seal, and that ten fingers and ten toes have twenty different names."

To communicate at all with the Inuit, the team members resorted to the usual gibberish and pantomime. "Some people relied on signs and drawings," wrote Lindsay, "others baby talk or English spoken slowly with an accent—any good foreign accent. One man was heard to use a word which he admitted was Swahili, 'but seems to do all right.'" (If these efforts indeed communicated at all, chalk it up to the skill of some of the Inuit to pick up a little English.)

A consequence of this linguistic impasse, though, was that the BAARE men remained completely ignorant of virtually all aspects of Tunumiut* religion, history, morality, and cosmology—all the things that constitute a culture at the deepest level. (Of this incomprehension, and what the team missed, more below, in "Interlude: The Cosmos of the East Coast Inuit.")

---

* The Inuit who live on Greenland's East Coast are known as the Tunumiut.

On August 13, the day after the *Quest* departed on the Northern Journey, an event occurred that at first quite flummoxed Percy Lemon. As Chapman tells the story,

> Three girls, beautifully dressed but rather shy and giggling, arrived at the Base and offered to clean the place up. Lemon got them to scrub the floor of the Base, to sew the curtains, and to do the cooking, while the men, who soon arrived in their kayaks, carried wood, and did other heavier jobs. The girls wanted to stay, but as food was short, and Lemon didn't want to compromise himself—chaperons being rather scarce in this district—he set off in a motor-boat to take them home.

Unfortunately, four miles out, the motor broke down. Lemon had to row the boat all the way back to base camp, with the "girls" still aboard. At once they installed themselves in the hut, choosing crannies in the attic above the main room in which to sleep. Without further demurral, Lemon accepted the bargain—as did, apparently, nearly all the men as soon as they returned from the Northern Journey and the trek to establish the Ice Cap Station. Among the other services the newcomers provided was to give the men haircuts.

Chapman's jaunty account of this new ménage imports all the unconscious assumptions of upper-middle-class British superiority, as well as the colonial notion that "natives" were happy to serve their "masters" as domestic servants. (The names by which the men addressed these Inuit women—and one man—were no doubt awkward stabs at their "unpronounceable" real names.)

> Our permanent staff consisted of Arpika, the oldest and most sensible of the girls; Gertrude [or Gitrude], the prettiest, who expected to be made a fuss of and at first only worked when she felt like it; and Tina,

Gertrude's younger sister, an incredibly sluttish and dirty girl who was given such jobs as cleaning the pans, peeling potatoes, and washing up. The party was completed by Gustari [or Gustare], a youth of about 18, who helped with heavier jobs such as carrying up coals for the kitchen stove, and ice for our supply of drinking water. Very soon he became extraordinarily efficient, learning to run the outboard motor, and even to start charging the engine on his own.

Likewise Lindsay, whose vignettes of hut life feature Arpika "showing off to an admiring circle the evening dress that she has made from an old pair of curtains"; Gitrude, who is "a sex conscious young woman, highly strung and inclined to be tiresome"; and Tina, "who was once seen blowing her nose on someone's sleeping-bag, and when she first came she had a habit of spitting on the dirty plates and then rubbing them with her fingers, so she was usually known to us as the 'Little Slut.' "

Except for Gustari, described as "about 18," it's hard to judge the age of these Inuit housekeepers. In the photos the men took of them, for which they happily posed, none of the "girls" looks older than eighteen. They could well have been younger—especially Tina.

What was in it for these women to become domestics for the *Kratouna*? They shared the meals, full of novel foods for them, that they cooked for the men. They were each paid a "salary" of two cigarettes a day. And there was the gramophone. Given free rein of the record collection, Gitrude played one in particular so often that it drove Lindsay to distraction. At last he seized the record, smashed it to pieces, and threw it away. Gitrude retrieved the pieces, pasted them back together, and tried to play the record one more time.

The men of BAARE started to expect the kinds of diligence and reliability that noblemen in Britain would demand of their servants. As Lindsay complained,

Although the staff always did their best, they needed a lot of supervision. They would never tell each other to do anything, so among themselves there was little organisation. If more wood was wanted in the kitchen, it did not occur to Gitrude to tell Tina to go and fetch another bundle. Instead she would wait until the fire had gone out, and then come to whoever was doing the catering that week and say, "There is no wood." He would then shout to Gustare to go and get more, and Gustare would run off with the best will in the world; but by the time that the wood was chopped up and the fire relit, half an hour had been wasted.

Chapman had kindred complaints:

The Eskimos had somewhat specialized ideas of cleanliness. When they were told to wash up the cups and saucers they threw them out of the window . . . and the dogs licked them clean. The natives were quite hurt when told to wash them again. . . . The kitchen stove would have to be relit six or seven times each morning, while coal and water were invariably "perangera" ["there isn't any"]. . . . Lemon always had to be very careful not to hurt their feelings, for they were very prone to sulk, weep, and have hysterical fits.

As the autumn months wore on, however, the relations of the four "domestics" with their hosts grew more complex—and more troubling.

*   *   *

As he headed out from the weather-monitoring station on October 5 with Jamie Scott, his best friend and most trusted partner in the BAARE, on the push across unknown ice cap that he would call the Southern Journey, Gino had every reason to believe the complex expedition he had devised—even with the compromised goals of the Northern Journey—was a success. Despite the nastiness of Buggery Bank, two parties had covered the 130-mile trek to the ice station without a serious setback. One pair of men had stood duty in the domed tent for five weeks, taken their readings, and suffered only minimally while finding much to please them in their novel isolation. Another team, having just taken over, should likewise flourish. The route from base to station was well-flagged and comparatively easy to follow both coming and going.

It was true that the wireless set, which would make the ice station more secure and its monitors less anxious, had yet to be carried up to the distant site, but Gino had hopes that radio transmissions might soon link the two vital outposts of the BAARE. And he had even grander plans: as soon as the Gypsy Moths were fully operational, one of the team's several pilots ought to be able to resupply the station via airdropped supplies. Gino even thought it possible that a plane could land there, which would render the relief of one party and the installation of another almost a piece of cake. At the beginning of October, morale among all fourteen members seemed sky-high.

During that first week of October, however, things started to go wrong.

Gino's plan was to sledge for six weeks, first reaching Nansen's traverse route on a south-southwest vector, then turning back to head east-northeast along the hypotenuse of a triangle toward base camp, hoping to intersect the established route somewhere near Big Flag Depot, but discovering a whole new massif of coastal mountains along the way. He was also prepared, if need be, to find a new route through one of the glaciers spilling off the ice cap, down to the coast somewhere south of base camp, then along the coast back to base.

Total distance: something like five hundred miles. If it took all six weeks, the pair would have to average about twelve miles a day. And that allowed for no stationary pauses to wait out storms. It was a bold plan, but the kind of challenge Gino relished, and the kind of exploring, stripped to a single partner and a minimum of supplies, that he lived for.

The best discovery the two could make would be that of a deep hidden valley or hollow at much lower altitude than the rest of the ice cap. That might then serve as the corridor for the trans-Greenland flights in the Gypsy Moths the team hoped to attempt the next spring, paving a path for the great British Air Route to western Canada that was the express purpose of the whole expedition. But aside from practical considerations, the two-man jaunt would embody Gino's exploratory ideal, to go where no one else had ever been.

Scott and Watkins started out on October 5 with two sledges hauled by seven huskies each, carrying a total load of 700 pounds, which Gino had calculated as full rations for both men and dogs. The start was smooth, on "old wind-drifts covered by 6 inches of powder snow," but even so, the men at first averaged only eight miles a day.

For five days, the temperature dropped steadily, reaching a new expedition low of minus 36 Fahrenheit on October 10. Despite the difficulty of staying warm, the men welcomed the cold with its better sledging conditions. But after that date, the temperature started to rise again, and a new surface gave them fits. Gino switched from skis to snowshoes, but the sledges kept overturning as they tried to breach the sastrugi (from the Russian for "small ridges," for the sharp parallel fins of hard-packed snow created by the prevailing winds, the bane of polar explorers throughout the centuries). Each time a sledge overturned, "the two of us wasted a lot of time and energy in righting them again."

Then a warm snowstorm hit the plateau, which succeeded in "filling up the hollows between the ridges," but it was followed by a plunge in temperature of no fewer than sixty degrees in a single night. The effect on the dogs was dramatic: all fourteen now became "listless and miserable." Gino fed the huskies extra food by cutting down on his and Scott's own meals.

Always one to see the bright side even of privation, Gino took heart in the fact that each day the sledge-loads were lightened by twenty pounds—the weight of the food consumed by fourteen dogs and two men. On the evening of October 15, as Scott later recounted, "We crawled into the tent as excited as if we were sitting down to a very special dinner." Burning paraffin, the Primus gave off fumes, as it had each night. Scott tended the pot, while Gino, whose eyes were sensitive to smoke, lay prone against the tent wall.

I knelt upright, stirring a well-filled bowl of pemmican, sprinkling in porridge oats and pea flour to make it still more appetising. Then, after no conscious interval, I was gazing at the stars and my head was full of noise. I did not know where I was, nor greatly cared. Gradually I realized that I was lying with my feet in the tent and my head and body in the snow outside, while the noise was partly singing in my ears and partly the yapping of a hysterical dog, puzzled by my behaviour.

Slowly regaining his senses, Scott crawled back inside the tent, where he found Gino "wringing my carefully prepared pemmican soup out of the light down sleeping-bag he used."

"Are you all right?" Gino asked. "At first I thought you had gone mad and wondered how I should get you home, and then I thought you were dead."

"What happened?"

"You suddenly blew your nose into the butter," Gino answered. "I thought that a bad sign; but when I protested mildly you started throwing your arms about and knocking everything over. So I turned off the primus and hauled you outside."

Scott had come close to being a victim of one of the most insidious perils faced by mountaineers and polar explorers camped in small tents. A poorly burning stove can give off carbon monoxide fumes, which are completely odorless and can quickly prove fatal. An inordinate number of adventurers have perished from this poisoning over the years, most without ever having an inkling of what hit them.

Because he was lying on the floor and close to the tent wall, Gino had mostly escaped the fumes, and his quick reaction saved the day. But in the time-honored tradition of treating near-disasters with deadpan irony, Scott later wrote, "Gino's strength and calmness had saved my life; but you cannot thank a man when he starts cursing you light-heartedly for making a filthy mess and ruining a good dinner."

For another two weeks, the men pushed on south, increasingly discouraged. Their own determination was almost as keen as when they had set out on October 5, but the dogs told another story. As Scott put it, they "seemed more bored than anything else: they wandered along without interest and sat down every fifty yards or so, while neither beatings and curses nor petting and encouragement could wake them from their lethargy."

Scott and Watkins realized by now that the goal of intersecting Nansen's route was out of reach. A trek of only a hundred miles seemed to Gino "insignificant," but he wisely determined to turn toward home when half the rations were gone. Even that strategy was cutting it thin, for the triangular hypotenuse back toward Big Flag Depot was a longer leg than the outbound march along the south-southwest vector. Accordingly, he called for a turnaround even before the team had covered the hundred miles.

Though neither man explicitly reached this conclusion, it seems clear that the dogs in their fatigue were attuned to dangers that Scott and Watkins, caught up in their goal-oriented push, were blind to. It would not be an exaggeration to say that in the end, the huskies saved the men's lives.

According to Scott, "As soon as [the dogs] had grown used to the new direction their tails went up and they began to trot. We did double figures for the first time and the next day Gino often had to ride on his sledge because he could not keep up with his dogs. He was disgusted: one week like this before we turned would have taken us to the line of Nansen's crossing and so fulfilled our main objective."

The huskies' new-found exuberance would lead, four days later, to the closest call of the whole Southern Journey. First, however, the team was engulfed in a violent windstorm blowing heavy snow for fifty hours straight. Outside the tent, the men "could scarcely breathe or see," but inside it they were warm, "content to sleep and talk."

While Scott and Watkins stayed cozy inside their sleeping bags, the dogs were suffering miserably in the cold and wind. They "rolled themselves up, their noses under their tails, and disappeared," Scott recorded. "Now and then one of them got up, leaning against the wind; he blinked the ice from his eyes and tried to shake the snow out of his coat. He gave it up, [then] stood disconsolately with his fur blowing up the wrong way."

After a fifty-hour layover during a fierce storm, the men packed up on October 29 and moved on. Both men wore snowshoes that alternated skimming the crust and breaking through, each extrication requiring a tiresome flounder back to the surface. In late afternoon, Scott stopped to adjust one of his snowshoe bindings. And as he did so, he made a serious mistake.

As he fiddled with his binding, he allowed the dogs to pull ahead with the sledge, "for it was only too easy to catch them up." Suddenly, however, the dogs crested a small pass. "The leader started to trot," Scott later wrote, "and the others followed him." He shouted "Unipok, stop!" again and again, but the huskies ignored him. Scott tried to run after them, but kept tripping in his snowshoes; after he removed them, the crust broke under the weight of his boots. The sledge pulled farther away. On it were lashed the tent, Scott's sleeping bag, and most of the food. Without those items, Scott admitted, "we would fare badly in another storm." In reality, without them, both men would almost certainly have perished.

Powerless to catch up, Scott yelled to his partner for help.

Gino took off his snowshoes and started to run. He ran lightly, with quick short steps; but he broke through every four or five yards. At first he gained rapidly; but as he began to grow tired the distance between him and my sledge remained much the same. Several times he made a sprint but just before he reached the handle-bars he stumbled through the crust on which the dogs ran easily. It began to grow dark. He knew that he must catch up soon or not at all.

Scott postholed along, watching the drama unfold ahead of him, only beginning to foresee the consequences of failure. But Gino "made a final effort, dived for the sledge, caught it and held on. The dogs, as if tired of the game, stopped and lay down at once."

When Scott stumbled up to his partner, Gino, still panting from the exertion, cursed, "Damn you, Jamie"—then turned the close call into a joke. "I'd taken a lot of trouble," he complained, "to stay cool all day and now I'll have to sit up half the night to dry my clothes." But after the expedition, Gino confessed—not to Scott or any of the other BAARE men, but to his father back in England—"that this had been one of the most anxious occasions of his life."

The storms that would have likely cost the men their lives, had Gino failed to recover the sledge, continued almost nonstop for the next six days, during which the pair advanced a paltry fourteen miles. In the tent, the men had trouble sleeping.

Gino treated the storm as a bad joke. "This is a damned silly way to spend one's time," he suddenly burst out. "We ought to be lying in our bunks with an Eskimo playing the gramophone and feeding us with seal meat and milk chocolate."

The storm finally died out. Though unable to take latitude readings with the theodolite, Gino and Scott estimated from the sledge-wheel count that they might be only thirty miles from the marked route to Big Flag Depot. Now Gino decreed that they should abandon one sledge and load all their essentials onto the other. Scott was impressed by how cavalierly his partner threw away belongings that were dear to him: "I knew this trait of a true vagabond, but still I was shocked to see his fine copy of the *History of England* lying in the snow with its pages flicking over in the wind."

On November 8 they entered a crevasse field, indicating the near edge of the ice cap. Gino took the lead and carefully wove a route, probing the snow bridges with a ski pole. Another day of whiteout dictated a rest; but early on November 10, an object to the side of the trail caught Gino's eye. It was a tiny scarlet thread. They must be very close to the flag-marked route leading back to base camp.

Only moments later, they saw "a broken black line which gradually resolved itself into men and dogs." By an astonishing coincidence, they had intersected Chapman's party that had set out on October 26 to relieve the Ice Cap Station. The two parties rushed to embrace each other. With Chapman were five teammates. They must be returning from their mission, Gino and Scott concluded, but as they looked into each other's iced-up faces, they realized that none of those six men was D'Aeth or Bingham.

With rising incredulity, Gino blurted out, "Where are you going? Not in to the ice-cap station?"

Chapman acknowledged the fact. His team of six had had an excruciating trek through the same storms that had battered Watkins and Scott. In fifteen days, they had covered only a little more than fifteen miles. Now it was too cold to stand around chatting, so Gino made a quick, grim calculation. He told Chapman, "Never mind about the wireless. Take its weight in food and concentrate entirely on getting in and bringing out Bingham and D'Aeth. You may have to abandon the station. I don't know. You'll just have to use your own judgment and do the best you can."

As the two teams parted ways, Gino was full of self-recrimination. *What had gone so wrong?* What, in the end, had the truncated Southern Journey accomplished? Scientifically, the trip achieved little, Scott admitted.

All the way back to base, Gino tried to figure out what fatal flaws had so soon threatened to wreck the BAARE. Perhaps, with his seven different mini-expeditions, each with its exploratory goal, he had stretched the team too thin. From a perspective ninety years later, we can say that the BAARE, like all the explorers of the ice cap before and after, had simply underestimated the difficulties that otherworldly frozen void, stretching beyond the horizons, with its violent and unpredictable storms, threw in the faces of all intruders.

As he and Scott made their weary descent through Buggery Bank, Gino was preoccupied by a single thought: how to save what was left of the BAARE.

* * *

In the base camp hut, by November the relations between the three Inuit "girls" and their British hosts had taken on a new dimension. The men's accounts—Chapman's *Northern Lights*, Lindsay's *Those Greenland Days*, Scott's *Gino Watkins*—are silent about this development. Lindsay alone hints at it very obliquely, with Victorian squeamishness, when he writes, "We systematically spoiled the girls who worked in the house. . . . We treated them exactly like children; but we brought them up in the way that no children should go."

It is only in the men's diaries, never intended for publication, that the more troublesome reality surfaces. At some point that autumn, Lemon started crawling up into the attic at night to have sex with Arpika. Not long after, Gino followed suit with Tina, the youngest, the "Little Slut." Finally Chapman got into the act, pairing off with "sex-conscious" Gitrude. The nightly fornications, noisy or muted, became a fact of base-camp life.

Understandably, this caused the first major disruption in the team's moral solidarity—though almost no hint of that friction crept into the "official" accounts. Quintin Riley, with his ascetic Christian bent, was deeply disturbed by the liaisons, especially that of his erstwhile best friend with an Inuit teenager. Lindsay too was upset. Tina's bunk was directly above his, and for weeks into the winter, when he was not off on exploratory missions, Gino spent every night upstairs. Privately Lindsay griped, "Every time this bloody Eskimo girl got up into his bunk she had to put her feet within an inch or two of my face. I objected strongly, and so did some of the others, and if the ship hadn't come when it did [in the summer of 1931] we'd have formed up and said look here, Gino, this has bloody well got to stop."

Lindsay's dudgeon explains a sentence in *Those Greenland Days* that otherwise hangs like a conundrum: "The female servants of an expedition should be as prescribed for bedmakers in an old University statute—*horrida et senex* [ugly and old]."

The reactions of most of the other team members remain unknown. Perhaps they simply shrugged, or perhaps they thought that since Gino was in charge and had indulged in the temptation himself, it was not for them to protest.

In the context of 2022, the couplings of Lemon, Watkins, and Chapman with Inuit teenagers could certainly be excoriated as sexual abuse, or even, depending on the true ages of the "girls," as rape. But there is another context in which those liaisons must be viewed: that of Inuit mores. By 1930 the Danish authorities were fully aware of the hazards of sex between white explorers and Inuit natives. That's why, on arriving at Angmagssalik, the BAARE men had had to bare their genitals for inspection for venereal diseases, and in Copenhagen Gino had had to lay down a bond in anticipation of any pregnancies that might result from such matings.

During the period when Arpika, Gitrude, and Tina were sharing the attic with their paramours, many other Inuit men and women, but mostly men, also visited base camp. They were curious about everything to do with these white-skinned strangers who had moved uninvited into their fjord. They seemed happy to perform all kinds of chores for the *Kratouna*, in exchange for dinners featuring unfamiliar foods such as margarine and sugar and coffee and biscuits. If it was too late to head back to their tents, they simply lay down in clumps on the floor of the hut to sleep, or climbed into the attic and fit themselves into whatever space was left.

We have, of course, only the Englishmen's side of the story, but there is little evidence that any of the Inuit men were angry about the sexual liaisons between the three teenagers and three of the explorers who had come from so far away. To probe this disjunction between cultural norms, it's worth looking at the experience of Europeans whose comprehension of the Inuit way of life far exceeded that of any of Gino's team.

Peter Freuchen was a Danish explorer who first went to Greenland in 1906, at the age of twenty-two. He fell in love with the country and the people, and ended up spending decades on the great island, accomplishing a number of dangerous expeditions, including the first ice cap traverse after Nansen's. In 1910, with his best friend, the great Danish-Inuit explorer and ethnographer Knud Rasmussen, he founded a trading station they called Thule, at 76° N. on Greenland's remote northwest coast, where the most isolated of all the Inuit bands flourished. There he became infatuated with Mequpaluk, a twenty-one-year-old Inuit beauty, whom he spotted around the trading post. He courted her for weeks with little success.

But one day she arrived to serve as a chaperone for Rasmussen's wife (also Inuit) in the house the two traders shared, while Rasmussen, as was his bent, took off on a solo expedition. Freuchen bided his time until an evening when the other woman was absent. Abruptly he told Mequpaluk "that she had better stay with me."

She looked at me a moment and then remarked simply: "I am unable to make any decisions, being merely a weak little girl. It is for you to decide that."

But her eyes were eloquent, and spoke the language every girl knows regardless of race or clime.

I only asked her to move from the opposite side of the room over to mine—that was all the wedding necessary in this land of the innocents.

This seduction sounds like something out of eighteenth-century Restoration comedy, translated to the Arctic. But among the Inuit in Greenland, a vastly different set of social norms from those of Georgian England (or 1930s Denmark) obtained. As other ethnographers have verified, simply sleeping together for the first time could be tantamount among the natives to getting married. But marriage itself in no way precluded other sexual pairings.

Freuchen himself was twenty-five when he married Mequpaluk, who promptly changed her name (as was the custom) to Navarana. For ten years, they shared an adventurous life, as Navarana joined her husband on fox-trapping and narwhal-hunting forays. They had two children together. The daughter became a well-known writer, and a grandson later served as the first Inuit member of parliament in Canada.

"She gave me some of the happiest years of my life," Freuchen wrote of Navarana in 1961. Yet even before they had met, the Danish explorer had learned about Inuit sexual customs that bore little resemblance to the amatory practices of Europeans. Lending another man your wife for the night had nothing to do with adultery or infidelity: it was a ritualized favor that was part of the fabric of Inuit society, and to refuse was considered an insult by both the husband and wife. Before he met Mequpaluk, Freuchen himself had engaged in the practice.

This casualness about sex did not preclude intense jealousy, however, and Inuit lore is rich with tales of murderous deeds committed by lovers driven mad with rage. This apparent paradox was one of many that somehow held communal life together generation after generation.

In the course of his many years in Greenland, Freuchen became completely fluent in the native language, and he came to understand Inuit culture better than any European before him—except his comrade, Rasmussen. His 1961 masterpiece, *Book of the Eskimos*, blends personal accounts of expeditions and the trading life with a de facto ethnography of the northwest Greenland Inuit. When it was published, the chapters about marital and sexual mores shocked European and American readers, in much the same way that Margaret Mead's *Coming of Age in Samoa* had in 1928. As Freuchen wrote,

Among the Eskimos, sexual life is not directly connected with marriage, and the simple biological need for the opposite sex is recognized in both men and women, young and old. Toddlers of both sexes are encouraged to play together with a freedom that would outrage a mother in America, and the game of "playing house" can—among Eskimo children—assume an awfully realistic appearance.

The sharing of wives was for the Inuit often a way of remedying threats to their very existence. If in a given season the hunting was bad, the game scarce, the resident *angakok* might order a general exchange of wives. He might even decree which wife should pair with which husband not her own.

There was also the rather popular game of "doused lights." The rules were simple. Many people gathered in a house, all of them completely nude. Then the lights were extinguished, and darkness reigned. Nobody was allowed to say anything, and all changed places continually. At a certain signal each man grabbed the nearest woman. After a while, the lights were put on again, and now innumerable jokes could be made over the theme, "I knew all the time who you were because—"

All these practices, evolved over the centuries by an intensely communal people for whom Western notions of "privacy" were unfathomable, lay at the farthest extreme from the hippie utopia of "free love" that flared into brief life in Europe and America in the late 1960s and early '70s. Nor did the Inuit casualness about sex preclude deep devotion between man and wife. Peter Freuchen had ten fruitful and joyous years with Navarana until she died suddenly in 1921, a victim of the world-wide plague called the Spanish flu.

Over his years in Greenland, Freuchen had formed a deep antipathy for the European missionaries who had so transformed (and, he believed, corrupted) native culture. Now his grief flamed into rage when the minister in Upernavik told him that Navarana could not be buried in the town cemetery because she had died a pagan, nor would he deliver a sermon in her memory.

"It was relaxing for me to be so furious," Freuchen later wrote. "I told him to go to the devil with his beliefs and his sermons, but my wife would sleep in the cemetery and not be thrown to the dogs. . . . I am glad that I did not strike him. I had the good grace to tell him to get out and let me manage the service." With a few helpers, he dug the grave himself and laid his beloved into it. A small crowd of Inuit watched from behind houses and rocks, terrified of incurring the brimstone wrath of the minister for those who defied the teachings of the Bible and the dictates of the church.

*   *   *

Whatever the actual nature of the relationships between the explorers and the "girls" may have been, there's good evidence that Lemon fell in love with Arpika and that Chapman did so with Gitrude. One entry from Chapman's diary hints at jealous passion on both sides of the affair, while it is at the same time hopelessly clouded by linguistic con-fusion. "Gertrude thought I said girls were beautiful in England," Chap-man wrote. "Actually I said I had no girl in England. She wept all afternoon, bless her. I cheered her up but this lingo is hell. She really is a charmer."

Lemon's affair with Arpika lasted through most of the year the team spent in Greenland. Near the end of his service with the BAARE, he learned that Arpika had somehow been carrying on a liaison with Knud

Rasmussen at the same time. Lemon was deeply distressed by the discovery of his lover's "infidelity."

Chapman's diary is full of endearments for Gitrude and of anguish over her endlessly shifting moods. His biographer, Ralph Barker, argues that the Englishman and the Inuit woman had fallen deeply in love with each other.

Gitrude subsequently married an Inuit man. "I adore her still," Chapman told his diary a couple of years later. "I shall never find anyone quite like her again."

What went on between Gino and Tina remains unknowable. If he wrote about the relationship in his diary, those passages have disappeared. Gifted, or cursed, by his extraordinary talent to compartmentalize the various parts of his life, Gino soon turned to what mattered most: the expedition itself, the endless allure of places where no one had been, and the frozen emptiness of the great ice cap.

The Cosmos of the East Coast Inuit

AS NAÏVE ABOUT other cultures as the members of the BAARE team may have been when they arrived in Greenland in July 1930, and as limited as they remained by the linguistic gulf between them and the Inuit natives, they were young men, and had the curiosity of youth, as many more seasoned explorers did not. (Sir John Franklin, ice-locked in the Canadian Arctic after 1845, let all 128 of his crew and himself die of scurvy, hunger, and hypothermia rather than learn how to survive from the Inuit whom he met near King William Island.) It was not enough for Gino's men to limit their grasp of the "Eskimo" way of life to what they could discern when the locals visited base camp. As early as September 17, Lemon and Chapman took the motorboat six miles farther into the fjord to visit the nearest settlement, called "Nettui" by its inhabitants. They found the Inuit still living in their summer tents but already hard at work repairing the old stone and turf structure that would serve as their winter home.

Among the first natives the two men met was an "aged widow" whose name they construed as Potardina, with her substantial extended family all tucked inside a single tent. "The occupants had already gone to sleep," Chapman wrote, "but were delighted to see us and produced some cold cod and sugar from under their sleeping-bench." (The sugar presumably obtained at the trading store in Angmagssalik.) It was at this moment that Chapman first met Gitrude, who took the lead in introducing the visitors to her tentmates, young and old. The next day the men first met Arpika, the daughter of an "aged man" named Nicodemudgy, whom they would come to know as the paterfamilias of the whole settlement. During the coming months Nicodemudgy would sometimes serve as a guide for the explorers, and in their down time he regaled them with tales (conveyed somehow through the language impasse) of his days as a great hunter, during which he had slain some ninety-five polar bears in hand-to-paw combat. He claimed he remembered every one. "There was the bear we tracked for a day and a night," he would tell the Englishmen. "He broke my lance and I had to strangle him with seal-skin line."

The first evening at Nettui, Lemon and Chapman invited all the local kids to their own tent, where they managed to fit nine boys and girls along with themselves inside a shelter designed for two. "Abandoning all conversation," Chapman recorded, "[we] gave ourselves up to the delight of ship's biscuits and sickly cocoa."

The men spent two days at Nettui. With a keen eye for detail, Chapman described the ingenious construction of tents made entirely out of seal parts, as well as the raised wooden benches ranged inside as sleeping platforms. He admired the "windows" made of sewn-together seal intestines favored by the "fairly well-to-do" families, and marveled over the great skill it took for a woman to keep a soapstone lamp fueled with seal blubber working properly.

Both men were struck by Inuit hospitality and friendliness. "We were asked our names, all of them, and then our ages: the latter being counted up in tens and the remainder on fingers," Chapman wrote. "They seemed very jovial people, always ready to burst out laughing, and after the first few minutes showed no signs of shyness." The image of the Inuit as a "happy" people, always ready to laugh, and essentially childlike, was by 1930 a stereotype with a long European pedigree. Throughout their year on the east coast, the members of the BAARE never seriously questioned that perception, and almost never caught even the more obvious hints of a darker side, a more adult side, of Inuit life—let alone of a universe teeming with ideas about conflict and morality and the supernatural that would have reduced their idée fixe about happy, laughing children to a caricature.

In mid-December Martin Lindsay spent three days in another Inuit settlement within the fjord (this one twelve miles from the base camp hut), on an errand to buy sealskins. And a week after Christmas, Gino, Riley, Chapman, and Rymill embarked on a sledge trip to the main Inuit encampment, twenty miles away, for a more extended visit with the natives. Both Lindsay and Chapman left vivid accounts of those encounters. Unlike such know-it-all predecessors as Franklin, the BAARE men had open eyes and minds about the exquisite adaptation of the Inuit to one of the harshest inhabited regions on the globe. In particular, the four men on the second sojourn were eager to learn just how the natives hunted seals through the ice throughout the winter. Chapman describes with unfeigned awe the patient vigil of a pair of hunters over the small hole they had gouged in the ice, one poised to strike with his twenty-foot harpoon, the other lying prone, caparisoned in sealskin, murmuring bubbling sounds to attract the prey.

Lindsay waxed effusive about Inuit inventions that made that life possible—most notably the kayak, made entirely of pieces of seal except for the wooden framework, that perfected sea travel among the icebergs and floes. "There is . . . no flabbiness or shoddiness in the lives that they live," he mused. "Everything that they make, from their *kayaks* . . . to their gloves and shoes, is in itself a perfect piece of workmanship and could not be bettered by any of the accomplished processes of modern science."

Yet at the same time, Lindsay recognized how perilous Inuit life was, dependent utterly on success in the hunting of seals. Despite the language gap, he became cognizant of episodes in the past when whole villages had succumbed to starvation, not before "eating the corpses of the dead." Apparently the locals had no qualms about admitting that extreme survival stratagem to Englishmen for whom the very notion was abominable.

"They tell me here that human flesh tastes quite as nice as bear-meat," Lindsay jauntily reported; "but that you can always read in a person's face if he has eaten it." He gleefully witnessed the bloody orgy of a freshly killed seal being butchered on the spot. ("The infants cry to be put down and then crawl between the legs of their elders, at all costs to stick a tiny hand into the gory mess.") When the natives tested him by offering him a small chunk of the raw, frozen meat, he surprised them by chewing and swallowing it, then delivering the verdict "Not bad."

Gino and his three companions were willing to spend a couple of nights inside the Inuit stone hut, crawling into their thin sleeping bags laid on top of hard wooden benches, and Chapman reported that "we slept well in spite of the strange surroundings," though they were awakened now and then by crying babies, and once by "a terrific dog-fight in the middle of the floor." As soon as they had entered the house for the first time, they were swarmed over by women who pulled off their boots, socks, and coats to dry, "and would have taken all our clothes had we allowed it." Without batting a verbal eyelash, Chapman added, "The winter-house is kept so warm that the natives usually wear nothing except a small loin-cloth, and we naturally followed their example."

This openness to Inuit culture, the men's willingness to suspend their British prejudices about decorum, privacy, and even morality, was spearheaded by Gino, who more than almost any explorer before him was determined to learn new skills for Arctic survival by going to school under the natives who had mastered it.

Yet for all this openness, the Englishmen could not suppress a gut-level squeamishness that vitiated their nonjudgmental pretensions. As he crawled for the first time into the fully occupied stone house in the main settlement, Chapman had to admit, "What struck me almost tangibly on entering was the appalling mixed smell of rotten seal-meat, urine, dogs and children, which was utterly nauseating when encountered for the first time." Lindsay fell back on arch Victorianisms as he observed that "the Eskimos have one or two habits that are not above reproach. For instance, a mother eats the parasites [lice] that she picks out of her children's hair." And:

> They are not lewd, but at the same time they make no scruple to allow themselves free rein in conversation that can only be called lavatorial. They will talk about any subject with delightful naïveté. Last night I was asked to admire a lady's breasts. I said, "Lovely!"—and felt rather an idiot.

Yet no matter how hard the men strove to accept Inuit life on its own terms, they never shook the caricature passed down by centuries of European voyagers among the Inuit, of a joyful, laughing, essentially childlike people. Having earnestly praised the natives' communalism, where "every act in the day of an Eskimo is one by which all the others in the brotherhood receive some benefit," Lindsay lapses back into the trite cliché: "So they all live joyfully together, one big happy family, and laughter runs like a bubbling brook all the day long." After dancing outdoors with an Inuit family, then playing "round games" with them inside the hut, Chapman concluded, "The Eskimos are a most carefree people, and on an occasion like this they go almost wild with delight, young and old shouting and laughing like happy children."

Only once does Chapman get a true glimpse of the deeper culture that made the Inuit cosmos comprehensible. One evening inside the stone house, as the children crowded around the visitors "as if we were exhibits in a zoo," one of the old men (jealous of the attention paid the strangers, Chapman assumes) "started a rival exhibition by giving us an example of the old-fashioned drum dance."

He beat a small seal-skin drum, bobbed up and down and sang a weird dirge, all at the same time. The others watched his facial contortions and listened to his lament half in awe, half-humorously. Now it is only a winter evening's amusement, a survival, a burlesque.

But fifty years ago, before the first white men came, the old men told him (or so he believed), the drum dance had "a deeper significance. Then the witch-doctor, called Angakok, used to beat the drum when he went into a trance and paid a visit to the moon, or to one of the other worlds. All quarrels were settled in this way. . . . But already this custom is half-forgotten among the younger generation."

This is the only mention of the *angakok* (the usual European spelling; the best Inuit rendering is "angakkoq") in any of the expedition narratives from the BAARE, and that shaman is referenced as an extinct relic of the past. But in 1930, as in 1880, the *angakok* was still alive and well, central to Tunumiut life, the tranced communicator between the people and the invisible world of gods, souls, the future, and the universe.

Chapman just doesn't get it. He's like a tourist from Santa Fe or New York City sometime in the 1920s watching the Hopi snake dance, and coming away from it thinking it was a show put on to entertain Anglo visitors. All the while the Englishmen thought they were embracing Inuit culture with open hearts, they had no clue what was really going on.

* * *

Gustav Holm, the first European to reach Angmagssalik (in 1884), made a serious effort to study the culture of the Tunumiut, and his findings occupy many pages of the massive report he published after his return to Denmark. But he was not a trained ethnologist, and though he assiduously recorded everything he could, he was hamstrung by his own notions of the need to "civilize" a benighted, heathen people.

Twenty years would pass before the first real ethnologist, a Danish scholar named William Thalbitzer, who spent the year of 1905–06 in

Angmagssalik, conducted the research that came out in 1917 in his magisterial treatise *The Ammassalik Eskimo.*

Yet a profound change had taken place during those twenty years. It had taken only ten years after Holm's breakthrough for the Danish government to install a trading station in the village. That mission was led by Holm himself. Johan Petersen, Holm's interpreter, was made the colonial governor. With him came the first missionary, a man named Rüttel. Their combined impact on the religion and culture of the Inuit was profound. Some might call it catastrophic. But Holm himself saw only progress. "These two men have worked with rare energy, perseverance and patience for the civilisation of the natives," he blandly asserted. "Some of the people have now been christened, and murder, polygamy and other heathen practices are now rare, if not entirely abolished."

This was nonsense, but whether Holm was blinded by his own crystal ball or the Inuit deliberately kept the practices of which the Danes disapproved out of sight cannot be ascertained today. The fact remains, though, that Holm's report, nearsighted as it was, became the only record of Inuit life on the east coast before the advent and interference of Danish law and Christian evangelism.

In the spirit of the day, Holm equated studying the Inuit with gathering up as many "artifacts" as he could, to be carried back to Denmark and curated in museums, where they could be analyzed to reveal how a society worked. The nascent discipline of ethnography was all about collecting, all about *things*. Holm carted back to Copenhagen no fewer than 715 objects, ranging from throwing sticks to bone needles to skin bags to "meat and blubber forks."

But he also made a sincere effort to understand Inuit "social life," as he called it, publishing his analyses under such topics as "marriage," "morality," and "mourning."

Holm was the first to document basic ideas about death, the supernatural, and the universe among the east coast natives. Humans, his Angmagssalik informers told him, consisted of three parts: body, soul, and name. The soul was "quite small, no bigger than a finger or a hand." Yet it was crucial to life: "When the soul falls ill, the man falls ill also, and when the soul dies, the man dies also."

Holm probed deeply enough into Inuit thought to learn many of the concepts about creation and the cosmos that were as far away from Western notions as any belief system could be, and that seemed deeply alien to an educated European at the end of the nineteenth century. It was in this realm that the *angakok* ruled supreme.

An *angakok* could rob a man of his soul—presumably the soul of Life. It was then the shaman's job to divine, through incantations, where the soul had flown, and to try to restore it to the victim, who meanwhile had fallen gravely ill. After death, the soul came alive again, and took up residence in the sea or sky—both good places to be, though the sea was preferable. And yet Holm could categorically state, "The *angakut* [plural] . . . play no important part in social life."

In Holm, again and again, his Christian missionary zeal trumped his proto-ethnographic curiosity. If the highest good the Danes could offer the Eskimos was to "civilize" them, their pagan practices and beliefs had to be dismissed as barbaric superstition.

Despite the evidence before his eyes of the crucial role women played in holding the social fabric together, Holm made blatant generalizations about Inuit wives—bound perhaps by his own Danish notions of feminine duty and subservience. Thus he wrote, "Women have social importance only so far as they give rise to quarrels among men. Their position is hardly better than that of servants."

Mourning upon the death of a loved one, as Holm saw firsthand, was intense, marked by "howling and groaning and abstinence from a number of things." The closest relatives discarded all their old clothes, and destroyed or threw away everything belonging to the deceased. Among the Tunumiut, Holm was the first to document their deep horror of touching a dead body—a revulsion shared by Inuit bands as far away as Alaska. Yet despite all this evidence of agony over the loss of loved ones, Holm would declare that cases of "truly deep grief" were very rare among the Inuit in and around Angmagssalik.

Though the *Kavdlunaks*, the white men, had only first appeared in Angmagssalik with Holm, the Inuit already had their beliefs about these intruders. They had no souls, and they were the progeny of grotesque couplings between girls and dogs.

Holm learned some of the fundamentals of Tunumiut cosmology. Strangely enough, for a people dwelling only one degree south of the Arctic Circle, where the sun around the December solstice barely peeps over the southern horizon for three hours, but in summer dips behind the northern peaks for barely two, that life-assuring (feminine) heavenly body was much less important than the moon, which was masculine. The moon was the regular destination of voyages undertaken by the *angakut*. Jupiter was thought to be the mother of the sun, while Vega, the second-brightest star in the northern hemisphere after Arcturus, was an entirely benevolent spirit that humans could call upon for help.

From one informant, Holm learned the Inuit story of how the sun and the moon created the stars. It is a harrowing parable of incest and violence. The moon started having sex with his sister every night in her sleep. Perplexed, the sun covertly rubbed soot on her bedfellow's shoulders one night, and when she discovered that the seducer was the moon, she "cut off one of her breasts and tossed it to her brother, saying: 'As you seem so fond of me, eat me then!'"

This confrontation somehow led to both the moon and the sun lighting sticks to chase each other with. But as the moon's stick—a device used to knock ice off a kayak—kept fizzling out, he blew on it to build up the flame. The sparks flew off into the sky and became the stars.

Holm's discussion of the *angakut* is rich and nuanced, which makes his earlier dismissal of those shamans as "exorcists" all the more puzzling. Any man could become an *angakok*, Holm related, but the training took three to four years and involved privations and sacrifices. In a remote "abyss or cave," the novice had to rub stones for three days until a spirit emerged from a rock. This caused the acolyte to die amid "horrible torments," but he came back to life shortly after.

Holm witnessed firsthand several instances in which an *angakok* went into a trance and seemed to summon up spirit voices or cause his drum to float about the room. But instead of accepting the healing ceremony at face value, once it was over he cross-examined the *angakok* to try to make him confess what "tricks" he had employed in his séance.

*   *   *

Twenty years after Holm, Thalbitzer's relatively disinterested probe into that same culture dug much deeper. Again and again, as one reads between the lines, Thalbitzer's own study refutes and corrects Holm's. After 1894 and the installation of a missionary at Angmagssalik, Holm would complacently credit (as cited above) conversions to Christianity for reforming a heathen people all too fond of murder and polygamy. But more than a decade later, Thalbitzer found that of the 470 Inuit living in and around the village, only twenty to twenty-five had been baptized. "The old traditions were still alive," Thalbitzer asserted, "it was merely a question of finding them." Unlike Holm, Thalbitzer made no attempt to collect "artifacts," for he had learned that the people resented the loss of tools and weapons and spiritually charged objects such as amulets that they had gone to considerable trouble to craft.

There is a modern sensibility—that of a true cultural relativist—in Thalbitzer's delight as he notes that "Happily, at the time of my visit [in 1905–06], the inhabitants of Ammassalik were pagan and unable to read or write. Happily indeed, the art of writing was a novelty among them. The poetry and epic literature of their forefathers lived solely along the chords of their mother tongue,—rested solely on the art of memorizing. In the texts and melodies published here, we find an echo from the soul of the people." Full of delight, too, was Thalbitzer's discovery that the lore that both he and Holm recorded only mined the surface of Inuit belief and legend. In effect, Thalbitzer directly repudiated his renowned mentor when he celebrated "the magnitude and wealth of the spiritual life which animated the tribe before it became Christianized."

Yet Thalbitzer's initial response, after the 1905–06 stay in Angmagssalik, had been disillusionment at the changes "civilization" had wrought. As late as 1894, just as the new missionary started work at Angmagssalik, there were twelve *angakut* wielding their powers, a ratio of one for every thirty-four inhabitants. During the year he spent in the village a decade later, Thalbitzer could identify only five *angakut*, or one for every ninety-four natives in a slightly enlarged population.

Thalbitzer reported that only two of the five *angakut* vigorously resisted the missionary's threats and seductions. Those two had "stoutly refused to be baptized and displayed their contempt for the new-fangled doctrines. Why should their spirit-world, they said to themselves, be less true than that proclaimed by the missionaries[?]"

But in the long run, Thalbitzer thought, resistance was bound to be a losing battle: "It will be a small matter for the European priest to convince this handful of heathens that they ought to have Christian names and Christian souls (just as they already smoke Christian tobacco and Christian pipes). He has means enough to convert them. His chief weapons are: contempt and mockery of the heathen traditions."

Thalbitzer's first report, a short monograph he published in 1909, ends in an eloquent lament for what was doomed to be lost:

It is merely a question of time when the Eskimo culture on the isolated coast of Ammassalik will be levelled to the ground and the defenceless human flock up there will be just as denationalized and demoralized as many other primitive peoples have been by the complacent teaching of the Christian mission. Once more will a beautiful and interesting culture, widely different from the European culture, be blotted out of existence. It is a sad fact, and doubly sad when one considers that the idea of nationality has nowhere been so ardently preached and so readily accepted as in that quarter of the globe, which is the focus of the Christian mission.

It took more than a decade, as well as a second trip to Greenland in 1914, for Thalbitzer to figure out what was going on with the Angmags-salik Inuit under the assault of a Christian missionary. What at first looked like the erosion and disappearance of traditional pagan belief turned out to be a survival strategy: after 1894, the *angakut* performed their rites in much greater secrecy than before, while paying lip service to Christian ideas of God and heaven and right and wrong. This is not to deny that the missionary made sincere conversions among the people: after all, new luxuries such as tobacco and coffee and sugar were blessings brought by the same Danes who wanted to depose heathen "superstition." But rather than dwindle and ultimately vanish, the old religion of the ancestors went underground.

On his second visit, he was able to win enough of the natives' confidence to be allowed to witness many of those underground rituals. And thus he was able to probe far deeper than Holm into the all-important role of the *angakok*.

In his 1917 treatise, Thalbitzer presents the "magic" that the *angakok* performs without any skeptical commentary. He never tries to unravel the *angakok*'s performance by analyzing its "tricks," as Holm did—much less cross-examine the shaman about his purported subterfuges. And Thalbitzer was the first to ponder a special language that all the *angakut* lapsed into in their trances. "The words are not sheer abracadabra," he wrote, "but obsolete or metaphorically used words, a kind of inherited art language, which contributes in a high degree to the solemn and mystical character of the original gathering."

Thalbitzer also recognized, as Holm did not, the social importance of the "drum duels," in which, before an assembly of villagers, two men traded improvised chants as they each beat a drum, hurling taunts at each other and sometimes butting an opponent's face with one's chin. These were not mere occasional entertainments, but serious rhetoric marshaled to adjudicate grievances that whole factions of a tribe harbored against one another. A lasting drum duel could occupy its two antagonists through most of their adult lives.

During that year in Greenland, Thalbitzer made a detailed study of the East Coast dialect, one that holds up favorably today. With two colleagues, he made a collection of Inuit music and melodies, ranged in categories such as "kayak songs" and "theatrical drum songs." But the bulk of his original contribution, published in the second volume of *The Ammassalik Eskimo*, is a massive collection of tales and legends, only a few of which had been recorded by Holm. Thalbitzer's compendium amounts to one of the greatest anthologies of folklore that might otherwise have vanished ever rescued by ethnologists. Yet in the published work he recorded only a "small selected part" of the tales he had transcribed, "for if I included it all my book would exceed all bounds."

Even as the Inuit continued to trade at the store in Angmagssalik, and to welcome strangers to the coast like the fourteen Englishmen who arrived in 1930, the old pagan, animistic view of the world and the universe kept its hold on the people. The *angakut* cured maladies, not Danish nurses. The old taboos held sway, so that paddling an umiak past the treacherous Puisortok Glacier, no Inuit dared to talk, or laugh, or smoke, or look at the giant ice cliff, or say its name out loud. Women still lay with other wives' husbands, the terminally ill still threw themselves into the sea, and drum duels sometimes still spiraled into murder.

William Thalbitzer's magnum opus stands as both the detailed record of the survival of the ancient ways under Danish rule, and as a timeless memorial to the "beautiful and interesting culture" that so diverged from European ideas about how a society ought to work.

* * *

By 1930, though, sixteen years after Thalbitzer's last stay in and around Angmagssalik, how much had that picture changed? If the traditional culture of the Inuit still had deep roots, how was it that in a full year among the natives, the BAARE men seemed virtually oblivious to—or ignorant of—its existence? How could Freddy Chapman, watching an old man beat a drum and sing and dance, insist that a once-vital aspect of that culture had been reduced to "a winter evening's amusement," a "burlesque"?

One answer could be that perhaps Chapman was right. During those sixteen years, the deep traditions of East Coast belief might indeed have faded and become trivialized. Comparable collapses of native lifeways under the assault of Christianity, mandated schooling, and the dazzling variety of new things, new foods, and new customs are well-documented around the world.

It was part of Gino's plan for BAARE that the rations the men consumed over the year would be supplemented by the foods the Inuit lived on, and not just through purchases of seals killed by native hunters. As Chapman would recall years later, "As we could not afford to buy or transport enough food to last ourselves and the dogs through two [?] winters, we planned to study the Eskimo methods of hunting and to live off the country." The urge went beyond the practical: "The interest and excitement of the hunting, as well as the exercise gained, would help us through the long Arctic winter when the sun disappears for months altogether."

Even less merely pragmatic was the whole team's enthusiasm about learning to kayak Eskimo-style, about which Chapman later wrote, "Of the various skills that I have picked up in the course of a fairly adventurous life, the achievement of which I am most proud is that of learning to manage an Eskimo kayak."

To that extent, the BAARE men showed an admirable curiosity about Inuit lifestyle—as explorers such as Sir John Franklin had not.

But there were glaring limitations to that curiosity. The language gap notwithstanding, it's striking how little the men asked the natives about their beliefs about the environment in which they lived. Surely when they told the Inuit about their bold plan not only to go up onto the ice cap, but to build a station in the middle of it and man it through the winter, the locals must have voiced their horror and pleaded with warnings. There was a rich lore of stories among all the various Inuit bands scattered around the Greenland coasts about the ice cap, the dangerous "inland." But Gino's crew seemed to glean only the most superficial generalities. Martin Lindsay's only comment on the subject, in *Those Greenland Days*, recycled a dubious legend that probably came from a garbled caricature handed down by white traders. According to Lindsay,

> Although man had long been familiar with the coast of Greenland . . . nothing whatsoever was known about the interior. It was pictured as being intersected by ice-free valleys, where, said the Eskimos, many beautiful four-breasted virgins, green in colour, tended herds of reindeer that browsed on luxuriant vegetation.

(No other source reports this fanciful Arcadia in the middle of the icy wilderness.)

Before he went to Labrador with Scott in 1928, Gino boasted that he had read everything he could find about that land. One would have thought that he (and perhaps other members of the BAARE) might have done the equivalent reading before arriving in Greenland. But there is no evidence that Gino had read or was even aware of Thalbitzer's *The Ammassalik Eskimo*. The book would certainly have been on the shelves of the Scott Polar Research Institute, founded in 1920 by two of Gino's mentors, Raymond Priestley and Frank Debenham. And James Wordie, another mentor, who led two expeditions to East Greenland in the 1920s (the first of which included August Courtauld as a member), surely knew Thalbitzer backward and forward.

At any rate, the apparent ignorance of the BAARE men—and of Gino in particular—about the culture, religion, and folklore of the Tunumiut is surprising, even shocking. It was the one part of their homework for the journey that they seem not have bothered to complete. If one had only the members' accounts of the Greenland expedition—Lindsay's, Chapman's, and Scott's—to go by, one might wonder whether by 1930 Christianity and Danish governance had indeed obliterated traditional Inuit belief—whether its vestiges, as Chapman deduced about the old man's drum song, had become mere "winter evening's amusements."

We are rescued from that fallacy by the advent of another great ethnologist, a Frenchman named Paul-Émile Victor.

In 1934, at the age of twenty-seven, sponsored by the Trocadéro ethnographic museum in Paris, Victor arrived in Angmagssalik to spend a year with the Inuit. His three colleagues included Robert Gessain, a doctor and ethnologist who also would become an expert on the Greenland natives. During that year, Victor, who had a natural talent at picking up languages, became fluent in Tunumiut. His fascination with the people only whetted his desire to probe deeper into the culture, so in 1936 he returned—first making a complete traverse of the ice cap by dog sledge from west coast to east coast with Gessain and two other companions. On arriving in Angmagssalik from that marathon journey, instead of heading back to France, he decided to travel some one hundred miles north to the fjord of Kangerdlugssuatsiaq, where a family had "adopted" him and welcomed his installation in their stone house. Victor spent the next fourteen months there without any contact with other Europeans. In doing so, he realized the ethnographic ideal of "total immersion," as few anthropologists ever had (Bronislaw Malinowski in the Trobriand Islands in 1914–16 having set the mold). Victor's intention, as he put it, was to live "like an Eskimo among the Eskimo."

A gifted writer, Victor wrote a two-part memoir of his 1936–37 immersion, titled *Boréal et Banquise*. ("*Boréal*" means "northern," with Arctic connotations; "*Banquise*" means "ice floe.") Victor's subtitle was "Joy in the Night." The 496-page work, as readable as a good novel, combines diary entries with keen ethnographic observations. The book won a following in France not unlike the one that Margaret Mead commanded in America for her similar *Coming of Age in Samoa*, published a decade earlier.

Boréal, but not *Banquise*, was translated into English in 1939 as *My Eskimo Life*. As if the story of life with an Inuit family needed any further spark to popularize it, Victor unabashedly centered his narrative on his affair with a nineteen-year-old "beauty" named Doumidia. The dust jacket for the Simon & Schuster edition clunkily promised, "You will be glad to meet Doumidia, the most charming debutante of the season." (It seems all too characteristic of the gulf between British explorers and their French counterparts in matters of morals and propriety that Victor celebrated his liaison with his Inuit "mistress," while Lemon, Watkins, and Chapman were at pains to suppress any hints of theirs with Arpika, Gitrude, and Tina.)

Thus only three years after the BAARE finished its campaign in Greenland, a single open-minded anthropologist with a talent for languages was able to gain deep insights into Inuit beliefs about everything from creation and the cosmos to the unspoken conventions of daily life—a realm of understanding that remained beyond the reach of Gino and his men. And what Victor demonstrates at every turn of the page is that in 1936 the folklore, the religion, the mores of the east Greenland Inuit remained fundamentally intact, congruent with the culture they had possessed before the missionary and governor plunked themselves down in Angmagssalik in 1894—no matter how much their new life often required awkward reconciliations with Christian teachings and Danish law.

Victor learned all kinds of beliefs about the ice cap that bore no resemblance to a Shangri-La of beautiful green-skinned four-breasted virgins raising reindeer. All the stories peopled the great inland wilderness with monsters, many of them giants. Victor named two of the puppies sired by one of the dogs that had pulled his sledge across the ice cap Timertsit and Ekridi—a giant that carried an entire cooking kit attached to its jaw, and a hairy, naked man who walked on all fours, respectively, both ice cap–dwellers and enemies of the Inuit. There were other scary beings that inhabited the edges of the ice cap. The *toupidek*, created by an *angakok*, was often seen as a chimera with the skull of a man, the body of a seal, the hindquarters of a bear, and human hands. The *tsoutoup-iwa* was a man who lived in an ice grotto and crept out at night right up to the windows of an Inuit hut to threaten its inhabitants. The *poubik* was an ice monster embedded in an iceberg in the fjord, which came to life to "fall on houses and crush them to pieces." Most frightening of all were little men, spirit helpers to the *angakut*, who lived in a region called Timek, right in the middle of the ice cap. As Victor summarized the Inuit belief,

> They live on the flesh of wolves or dogs, and have large, prominent stomachs. Whenever they eat a man, they leave the skeleton intact; and the man's flesh, having been rejected later in the form of excrement, returns to the skeleton. Little by little the body then regains its old form, until finally the man comes to life again; whereupon he himself becomes an angakok.

As such tales reveal, the *angakok* was a person who provoked ambivalent feelings: he could use his powers to search for and restore lost souls to very ill men and women, but he could also bestow curses and hexes on his neighbors for reasons no one could understand.

One of the few pieces of Inuit lore about the ice cap that the BAARE men learned from the Inuit was their claim that polar bears, on their long migrations back to their hibernation dens, often headed far up onto the frozen plateau before cutting down to the coast from time to time to fatten up on seals, finding the going easier on the ice cap than along the intricate zigs and zags of the shoreline. Chapman, among others, chalked this claim up to sheer superstition, until his relief party came across a solitary polar-bear track in the snow at an altitude of 5,000 feet, many miles inland.

Victor's romance with Doumidia, as they lay together night after night on bunks in one corner of a single-room hut that measured only twenty by sixteen feet, with a ceiling only five and a half feet high, along with twenty other Inuit, the majority of them children, seems to have been accepted by all. Some of the scenes in which the author portrays his relationship with the nineteen-year-old vividly mirror the tension between the old pagan faith and the Christianity newly imposed upon the people. One day Victor read aloud, in French, from Ecclesiastes. Though they understood not a word, Doumidia and two others listened entranced. When he stopped, they begged him to continue. Perhaps perversely, Victor instead sang an Inuit chant he had learned, "one of their old pagan songs of days gone by."

"Be quiet," pleaded Doumidia. "Don't sing that." (All the conversation, of course, took place in Inuit.)

"Why not?" Victor answered. "These songs are your songs. We have our songs too."

"Yes, but God wouldn't like it."

Victor launched into a defense, half-ironic, to the effect that God knows everything and knows that what he is doing is not wrong.

Doumidia surprised him with her rejoinder. "How funny!" she said. "Now I see for the first time that you have thoughts. I believed that you hadn't got any, that you were just like the other Kratouna."

"What! You really believed that the Kratouna had no thoughts?"

Doumidia wouldn't back down. Moments later, she insisted, "I had always believed that we were the only people who really understood God."

Through Kara, Doumidia's forty-five-year-old mother, Victor

learned about the Great Hunger of 1882–83, when famine and starvation had decimated the tribe. The stories of that terrible time were still alive and painful more than fifty years later. It is very surprising that Holm made no mention of the Great Hunger, nor of the *angakok*'s prophecy that at the end of the tribulation, white men would come and lay waste to the survivors. Holm's sudden arrival in August 1884 ought to have signaled the fulfillment of the prophecy. The natives' first response must have been terror.

Instead, Holm insisted instead that as he sailed into the Angmagssalik harbor, the inhabitants greeted him with claps of astonishment and greetings of welcome. Nor does he anywhere hint at a terrible famine that ravaged the people during the two preceding years.

Was Holm simply dissembling? No explorer likes to go down in the record as the agent of doom for the people he "discovers." Or did the Inuit manage to hide all signs of the terror they felt, and avoid altogether talking about the bad years?

The tales of the Great Hunger that Victor learned from Kara were so distressing that he left them out of *Boréal et Banquise*. Only in 1953 did he bring them together in a slim volume titled *La Grande Faim*—published two years later in an English translation as *The Great Hunger*. Victor's method was not to write an objective account, in the pattern of Defoe's *A Journal of the Plague Year*, but simply to tell the stories through the mouths of their protagonists, as Kara (and one other oldster in the family) recited them.

The whole collection makes for grim but vivid reading, as men, women, and children are reduced to eating their dogs, the skins of their kayaks and umiaks, even their own boots. Children are forced to commit suicide by jumping off cliffs into the sea. The most terrible stories involve cannibalism, about which the Inuit developed an attitude of fatalistic practicality. One of the most harrowing passages recounts the story of Iguimadek, who with his wife undresses a girl who has just died, "cutting her up as if she were a seal." Iguimadek fills a basin with the girl's intestines and goes outside. As he walks through gently falling snow, he remembers her.

> They had grown up together. . . . He had played with her many a time among the rocks. . . . When they left the hut, she would set out a little ahead of him, and he would follow, his hands in his pockets. As soon as the house was hidden behind a hill, he would catch up with her and pull her plaits, and they would knock each other about for a minute or two. . . . He used to love the contact of her warm body in summer on the grass in the valleys, and in winter at night in the hut.
>
> The wind rose, penetrated the man's jacket. . . . "I'm cold and hungry," he thought.
>
> He walked all round the hut. Then he put down the basin on the roof, which was on a level with his knees, and began to eat its contents.

Victor had decidedly mixed feeling about Christianity. He deplored its impact on cultures that had coherent religions of their own. In Kangerdlugssuatsiaq Fjord, he saw the losses every day. Kristian, the best hunter in the band, told Victor that since he had been baptized, he had lost the ability to see monsters.

Boréal et Banquise is such a rich account of life with the Inuit that it's impossible to do justice to it here. Years later, though, Victor said, "Among my own people, I always felt that I was 'other.' Among the Eskimos, I felt that I was one of them. I learned how to be a person. I became a person."

Near the end of Victor's stay with his adopted family, Kristian, who in his own shy way had been courting Doumidia, approached him.

"What do you think?" the hunter asked the Frenchman.

"About what?"

"Uh—Doumidia and me."

"Well?"

"Do you think I could have her?"

Since Victor was approaching the time when he would leave his "family," he too had been worried about Doumidia. That the Inuit man thought he needed, in effect, to ask Victor's permission embarrassed him, but he thought that no one in the band would have made a better match for his lover. He gave Kristian his whole-hearted approval.

The hunter embraced the ethnologist, then said, "I shall go and ask her at once."

Doumidia and Kristian were married soon after. If Victor's account can be trusted, everyone was happy—not only the Inuit pair and himself, but everyone in the hut.

The abiding conclusion Victor came to after spending two full years with the Angmagssalik people was that, for all the inroads of Christianity, of the trading post, of Western goods and European visitors, the traditional Inuit culture was remarkably intact. It would not always be so. In 1969, after another extended visit to the east coast, Victor's longtime colleague Robert Gessain would write a sad, brilliant, bitter book called *Ammassalik, ou la Civilisation Obligatoire* ("obligatory" in the sense that the Inuit had finally been brainwashed by the missionaries and corrupted by their dependence on imported luxuries—in particular, alcohol, which in 1936 was still effectively banned). One characteristic chapter's title translates as "In place of seals to butcher . . . beer, aquavit, and cards."

*   *   *

It would be unfair to the men of the BAARE to judge their interactions with the Inuit by the standards of Paul-Émile Victor. Gino's team had come to Greenland for exploration and science, not for ethnography. To their credit, they got along well with the Inuit, and their memoirs recount not a single serious conflict with the indigenes. Whatever their first impressions of the natives, whatever their squeamishness about the smell of the huts or the casual nudity the Inuit practiced, all the men came away from their year in Greenland praising the "delightful people" who were their hosts. If that judgment rested on the old stereotype of the Inuit as a friendly race, always laughing and smiling, essentially childlike, at least that idée fixe imported no malice or contempt generated by colonial notions of who should rule over whom.

If the BAARE men remained oblivious to the deeper culture of the Inuit, at least that ignorance did no harm. It was merely their loss—both intellectually and emotionally.

Gino and his teammates never forgot that they were visitors. Time and again, they expressed their admiration for how brilliantly the Inuit had adapted, and over so long a time, to one of the harshest inhabited regions on the globe. It was not a way of life that Chapman or Lindsay or Scott or even Watkins would have traded for his own, but the men never got over their awe at its idiosyncratic perfection.

EIGHT

"All, All Alone":
Courtauld on the Ice Cap

Alone, alone, all, all alone,
Alone on a wide wide sea!
And never a saint took pity on
My soul in agony.

Coleridge, The Rime of the Ancient Mariner

THE FIRST RELIEF party to follow the flagged route onto the ice cap—the team of six, headed by Gino and Scott, supported by Chapman, Rymill, D'Aeth, and Bingham—had a straightforward journey of 130 miles all the way inland to the weather station. Having left base camp on September 21, Chapman's foursome caught up with Watkins and Scott, who had departed five days earlier but had changed their plans for the Southern Journey en route. With new ski bindings designed by Rymill, and eager, disciplined, well-fed dogs, the men averaged eighteen miles a day on the ice cap itself—so fast that, as Chapman wrote, "it was hard work trotting or walking beside the sledges."

That team of six arrived at the station on October 2. It had taken Chapman's contingent only twelve days to make the long trek. D'Aeth and Bingham replaced Riley and Lindsay for the next stint of station monitoring, though Chapman reported, "Riley and Lindsay were quite happy and quite loath to leave." All eight men crowded into the big domed tent for dinner, where the combined fumes from the Primus and the nonstop pipe-smoking made their eyes stream and gave them trouble breathing. At last the newcomers escaped to their smaller tents to sleep.

On October 4, Chapman, Rymill, Riley, and Lindsay left for base camp, while Watkins and Scott set out to the south to launch their Southern Journey, hoping to cover more than 200 miles to intersect the route of Nansen's 1888 traverse. Before taking off, Gino instructed Chapman to launch a second relief party as soon as he got back to base camp. The return journey turned out to be much harder than the men expected, as the thermometer plunged, storms and high winds trapped the men for several days, and new snowdrifts played havoc with the sledges. The same conditions during the same days were forcing Scott and Watkins drastically to curtail the Southern Journey. Chapman's foursome got back to base only on October 14. The downhill trek with much lighter loads had occupied only a single day fewer than the much harder outward journey.

Gino's parting orders to Chapman on October 4 were characteristic of his optimism and ambition. Chapman was to round up "as large a party as possible" for the second relief mission, and to "start off from base as soon as they could." Moreover, that party was charged with lugging a wireless set all 130 miles up to the domed tent. If a working radio could be installed at the Ice Cap Station, with regular communication to and from base camp, that would mean a huge boost in safety, morale, and logistical efficiency. Chapman's team was also supposed to carry enough rations to last the monitors for five months, until the end of March. The plan was to relieve D'Aeth and Bingham from their lonely duties, and to install Wilfred Hampton and Alfred Stephenson (called by their teammates "Ham" and "Steve," respectively) in their place. Both were Cambridge men, Stephenson having graduated only one month before the *Quest* left London. He was designated chief surveyor for the BAARE. Hampton was a pilot in the Air Force Reserve, so Gino took him on as the team's aircraft engineer. So unclouded was Gino's crystal ball on October 4, as he delivered instructions to Chapman, that he planned for D'Aeth, once returned to base camp, eventually to fly Chapman and Rymill in one of the Gypsy Moths up to the station, land there, and swap the men out for Stephenson and Hampton. This scheme presupposed getting the wireless set to the station to coordinate the plan.

During the first team's absence up on the ice cap, Courtauld had overseen the invaluable work of getting all the supplies for the second team hauled to the base of the glacier, above which Buggery Bank reared its daunting track. To save weight, all the pemmican for both men and dogs had been taken out of their metal cans and repacked into cloth bags sewn closed by Arpika, Gitrude, and Tina. As Chapman later wrote, "Subsequently this turned out to be a very mixed advantage." The wireless set presented a major problem. Its charging motor alone weighed 150 pounds. A sledge loaded with such heavy equipment was a sledge that could not carry food.

Taking Gino's orders literally, Chapman reasoned that "the size of [the party] was only limited by the number of dog-teams available." Accordingly, he "enlisted" a team of dogs from Angmagssalik, "even though some of them looked as if they had escaped from a circus." The Tunumiut had never bred and trained their own huskies for long-distance sledging, which was why, before the expedition, Jamie Scott had traveled to the west coast to buy fifty dogs for the BAARE.

Dogs, in fact, would be the main reason for a twelve-day delay in launching the second relief expedition. Chapman was the only one of the available men with any experience in dog-sledging, and his apprenticeship amounted to the first relief journey just completed. Three of the available teams among the dogs still confined at base had never been driven since Scott had acquired them half a year before. Now simply rounding them up turned into an exasperating trial. "In some cases," Chapman reported, "we had to drive them on to the end of a point and even then they would walk out into the water rather than submit to being caught." Meanwhile, the days were growing colder and shorter, and a couple of blizzards stymied preparations.

It was not until October 26 that nine men started up the glacier: Chapman, Lemon, Courtauld, Wager, Stephenson, and Hampton in the relief party, with Riley, Lindsay, and Iliffe Cozens (the team's backup pilot to D'Aeth, in charge of aerial photography) in support. Chapman hoped those three could help the team surmount Buggery Bank and go as far as the Big Flag depot, fifteen miles from base, before turning back.

The team got almost nowhere during the first three days of brutal effort. The blizzards had scoured the whole lower glacier, not just the Bank, down to bare ice. Even pitching a tent became a dicey business. During the first storm, Lemon and Chapman anchored their shelter (pitched on a steep slope that plunged toward a frozen waterfall above a "rocky gorge") with a metal spike driven into the ice and 600 pounds of gear weighing down the flaps. Lying in their sleeping bags, they listened to the storm rage as night fell. Then:

> We dressed and put on our boots, knowing the tent could not stay much longer. Several times the tent-opening blew undone and at about 11 p.m. the end came. After a gust of prodigious force the outer cover was whirled up on one side, and soon blew away, scattering boxes over the edge as it went. We hung on to the poles of the inner tent for a few seconds, being lifted bodily off the ground. Fearing it would carry us down the glacier we let it go. . . . We lay in our reindeer-skin bags for a few seconds wondering if it would be better to risk being blown down the glacier inside the bags or to try to crawl to the other tents.

Chapman and Lemon chose the latter course, jamming themselves inside Hampton and Courtauld's tent, where the space was so constricted that "we had to turn over by numbers, all facing the same way at the same time."

At dawn the wind abruptly stopped. The men crawled out of their lairs to assess the wreckage. The charging motor for the wireless had been blown several hundred yards down the slope, but otherwise the damage was "surprisingly little." Later that day a search party recovered the missing tent in the rocky gorge below, still intact and usable.

It was only on October 30, the team's third day out of base, that they came to grips with Buggery Bank. There was no hope that the dogs could pull the sledges up the sharp-edged ice, so Chapman organized a block-and-tackle apparatus and the men winched the sledges up by brute force. The effort, Chapman swore, was "more exhausting than anything I've ever done in my life."

The storm had plastered the men's faces with rime ice. "My eyes were frozen solid," reported Chapman, "so I compromised and kept only one eye open." Wager's beard had congealed into "a mass of ice several pounds in weight." Stephenson solved the problem by cutting off Wager's beard altogether.

So it went, day after demoralizing day. During the first storm, two of Courtauld's dogs had escaped. Now, on the relatively safe slope above the Bank, Lemon left his dogs unattended for a few minutes, only to return to discover that the huskies had torn open a bag of pemmican and scarfed down twenty-five pounds of the dense ration.

Above Buggery Bank, the going was easier, but treacherous, as the team wove its way through a huge crevasse field. Sledge-loads had to be divided, requiring two trips each to get the full cargo a hard-won mile or two forward. Wager and Stephenson roped up and led through the crevasses, probing with their ice axes, but a number of men each stuck a foot through a snow bridge hiding a crevasse before jerking back in fear. Sastrugi regularly overturned the sledges, and the storms continued almost unabated.

On November 2, eight days out, the supporting trio of Riley, Lindsay, and Cozens turned back, still short of Big Flag depot. Trying to anticipate the worst, Chapman had put the team on half-rations early on, for fear the journey to the Ice Cap Station would take so long that the men would be forced to break into the ration boxes intended for the next pair of monitors. All the snafus, all the hardship, seemed to come crashing down on Chapman's spirit on November 5. In his diary, he complained, "The dogs are in a bad way. We hear them whimpering and many are at large. Things are incredibly uncomfortable. Our fur bags are soaked inside, possibly because we rarely have time or courage to undress. . . . When we wake up, the clothes round our faces are a mass of ice from our breath." Taking stock, he calculated that the team had traveled only ten miles in eleven days.

The situation forced Chapman to contemplate a grim bargain. To arrive at the Ice Cap Station without ruthlessly depleting the supplies meant for the coming winter months, he would have to reduce the team of six by half. And then push on, working even harder. But which three teammates should he send back to base? He delayed the decision for another four days.

On November 5, Chapman had discovered that his feet were numb, and their condition showed no change by the next day. Terrified of frostbite, he nonetheless kept his agony to himself. The storms kept up, almost without interruption. Excerpts from Chapman's diary during the next four days of fitful progress:

November 6: Our tents simply can't last much longer. The dogs have all bitten free [from their harnesses] and are huddled together against the tent for protection. . . .

In the evening the gale became ghastly. . . . We could not hear each other speak and hoar-frost poured over us like snow. The wind must be well over 100 m.p.h. . . . If the tent goes we are corpses.

November 7: Amazed still to be here. Lemon was jammed in between a ration box and a hard drift which has pushed in the side of the tent. I had to go outside and dig it out before he could move. . . .

Read *Tess of the d'Urbervilles*: a suitable book in this place; elemental strife in both.

November 8: Courtauld's dogs have eaten all their traces and several harnesses and most of his lash-line. In most cases the heat of the dogs' bodies lying on the drifts has embedded the traces into solid ice. The rope then gets cut to pieces as we hack them out. . . . Most of the dogs are limping and look half-dead. . . .

November 9: Gale all night and all to-day. My sleeping-bag has half an inch of water inside.

On the eighth, the team finally reached Big Flag Depot. Chapman left a note for Watkins in a tin can attached to a flag. Fifteen miles in fifteen days. . . . At that rate, they could never reach the station.

It was on November 10 that, by the rarest of coincidences, the team ran into Watkins and Scott arriving from the last leg of the Southern Journey. As Chapman recorded the meeting: "Suddenly saw a small moving object in the distance away to the left. A dog? A bear? . . . Went to meet them. They looked absurd, both sitting on top of a vast sledge with a whole galaxy of dogs in a fan before them." Wager: "Had we or they been fifteen minutes earlier or later we should not have met."

As related in chapter 7, on meeting the team of six, Gino first assumed they were on their way *back* from the Ice Cap Station. As soon as he learned the truth, he made a quick and desperate revision to the whole plan for the station, urging Chapman to dump the wireless, focus on bringing out Bingham and D'Aeth, and possibly abandoning the station.

In the cold, the meeting lasted less than an hour. But during that interval, Gino decided which three in Chapman's team should turn back. He chose Lemon, Stephenson, and Hampton, despite the fact that the last two were slated to relieve D'Aeth and Bingham as monitors through the coming weeks. In their stead, Wager and Courtauld were thrust into that role.

The parting was painful. Wrote Chapman, "I really felt quite sad to see them turn their sledges in the other direction and leave us." And: "We have arranged to have a dinner in London next Armistice Day."

Before separating from his companions, Chapman sorted out the best dogs to make three teams with three sledges, each pulled by seven huskies. And "We also took the best of everything they had, sledges, clothes, books, whips."

Then once more, Chapman, Wager, and Courtauld headed west-northwest, into the infinite blankness of the plateau.

* * *

Meanwhile, more than a hundred miles away, Bingham and D'Aeth kept up their chores at the Ice Cap Station. As it had for Riley and Lindsay during the first stint, the routine of six gauge-reading errands per day, interspersed with endless idle hours inside the domed tent, gave a sense of purpose to the men that kept at bay the eeriness, the psychic abyss, of total isolation in a white miasma ruled only by the ever-changing weather.

By November 10, when the meeting between Chapman's team and Scott and Watkins took place near Flag 56, Bingham and D'Aeth had manned their outpost since October 4. Both men expected their sentence to last only five weeks, perhaps because on parting, Gino might have promised as much. But by November 10, they had been at their work for thirty-eight days, or three days more than five weeks.

As early as October 30, Bingham had written in his diary, "Hope to see the relieving party some time about this day week." But November 5 came and went with no sign of sledgers in the distance, as did each of the following five days.

Now, as their service started to stretch beyond five weeks, Bingham and D'Aeth could not resist (just as Riley and Lindsay had not resisted) staring east along the flag-marked route each time they exited the tent, hoping to see those black specks in the distance. As watchers and wait- ers always try to do, each time the gaze revealed only the eternal snow and sky, the men pretended that it was no big deal to be disappointed— that a few more days of service at the station were nothing to complain about. Thus Bingham in his diary on November 14: "Walked a short way out along the flags to clear them. Judging by the intense cold on one's face caused by the low temperature I am very sorry for the relief party coming up. No wonder they are behind time."

The next day: "Rather a nice day with good visibility. Thought we saw the relieving party in the distance, but a mirage made it difficult. Possibly they may arrive tomorrow."

During their hours inside the tent, all month long the two men had read their books, smoked their pipes, cooked their meals, played cards, and talked for hours. They had also puttered around inside the tent, making little "improvements" to increase the homey comfort of their shelter. The biggest difference for Bingham and D'Aeth, compared to Lindsay and Riley, was the cold. Some of the trips outside to read the gauges, during high winds and lashing snow, were grim errands; and more and more, the men found that they were really warm inside the tent only when the Primus was firing or they lay tucked inside their sleeping bags.

After November 10, Bingham's diary entries recorded almost daily letdowns in their hopes for relief, even as he strove for stoic acceptance and refused to speculate what might have gone wrong.

November 16. We now examine the back trail with glasses when doing the observations, but with no result.

November 17. Good travelling day. . . .

November 18. Beastly day! High wind with no visibility. . . .

November 19. Nice day and we got a lot of clearing done outside. Still no sign of relief party arriving, so we should get the whole place cleared before their arrival. . . .

November 23. Left the Base ten weeks ago. Nice calm clear day but no sign of relieving party. Walked down the flags a short way to clear them. . . .

On November 15, the men recorded an evening temperature of 51.5 degrees below zero Fahrenheit—by far the coldest yet. The stoic tone of Bingham's diary may reflect only the brave face he was willing to show on paper. No doubt more and more of the talk the two men shared veered into theories of what might be causing the delay—like Lindsay's only half-joking scenarios about the base camp hut burning down or the whole relief party falling together into a crevasse.

It's human nature, in the absence of any real clues about a longed-for event, to plunge into explanatory theories, which tend to get wilder as the resolution is delayed. And as the waiting stretches on without an answer, the refugees often edge into anger, along the lines of, "What the hell are they doing that takes this long?" or "Have they simply forgotten about us?" There's no evidence that Lindsay, Riley, Bingham, or d'Aeth reached that depth of exasperation. Bingham's diary resolutely avoids blaming the relief party, or even speculating about disasters causing the delay. If D'Aeth kept a diary, it's either missing or not accessible. But such was the loyalty of the whole BAARE team toward one another during more than a year in Greenland that it's possible that the pairs of monitors never did begin to impugn the men who were so overdue in relieving them.

Instead, a new kind of anxiety started to take hold in late November. Bingham signaled it in his diary on the twenty-fifth:

Blowing a stiff gale outside. Everything drifting up fast. The entrance to our tunnel which we cleared this morning is already blocked again.

The tent is practically steady, but even with the snow house [protective walls] round it and a layer of drift snow in places feet thick all over it, the buffeting of the wind is quite alarming. Am not fancying the job of doing the 10 p.m. observations. Sorry for the fellows down the trail if they are getting this gale.

It was one thing when Lemon and Chapman's tent blew away down into the rocky gorge, for there were two other tents nearby for safety. And base camp was only a few miles away. But if the big domed tent at the weather station blew away, or collapsed, or the wind tore holes in it, Bingham and D'Aeth would instantly face a survival predicament against long odds. To build an emergency igloo with the gale raging, or a snow cave burrowed into the ice cap, might have proved beyond the powers of the pair. Even if they could have accomplished that feat, to hang on indefinitely awaiting the relief party might have posed a nearly impossible ordeal.

During the five days after November 25, both men spent hours outside the shelter simply digging out gauges and the exit tunnel itself, as the relentless wind and storm undid their every effort. On the thirtieth Bingham was fooled by an optical illusion that betrayed how desperate the men were for some connection with their teammates. "[T]hought the bits of paper on the snow about forty yards off were the relief party about two miles away," he admitted to his diary. The men were not out of food yet, partly because their stomachs had shrunk from inactivity and they ate less and less week by week. There were still two unopened ration boxes. "Thoroughly tired of pemmican," Bingham griped, "and hope the relief party have some meat of some sort."

December 1 marked the pair's fifty-ninth day at the Ice Cap Station: eight weeks and three days. By their reckoning, the relief party was now three and a half weeks overdue. There was nothing to do but shovel away the drifts, dig out the exit tunnel, read and record the gauges—and wait.

* * *

With their pared-down loads and the best dogs pulling three sledges, Chapman, Wager, and Courtauld started onward on November 11. It was on that very day that Courtauld first broached the idea of his manning the Ice Cap Station alone through the heart of winter. In his diary, Chapman summarized Courtauld's earnest proposal. Chapman's entry mingles his own fervent desire not to give up and abandon the station—the heart and soul of Gino's whole conception of the BAARE—with the skepticism of a leader responsible for his teammates:

> Courtauld says he is used to being alone and is very keen to try the experiment in such conditions. With so many books, a good supply of tobacco and ample food for one man, he says he will be perfectly happy. . . . In England few people objected strongly when Watkins said one of us might have to stay alone at the Ice Cap station in case of necessity, and this looks like the case in point. Anyhow, we can't decide anything yet: we've got to find the station first.

At first the three men made better time than all six had before the chance meeting with Watkins and Scott on November 10. But now a new complication threatened their progress: it became harder and harder to find the flags marking every half mile of the route, thanks to frost feathers occluding the red squares of cloth, the continuing storms, and the shorter hours of daylight. Chapman: "It is an awful strain gazing into the void looking for flags. You simply have no idea what focus to use as there is nothing to focus on. We thought we saw a big dark-coloured object in the distance to-day but found it was a small piece of black paper only 10 yards away." Later the men would conclude that some of the flags had collapsed and been buried under new snowdrifts.

The first evening, all three men crowded into one tent. But after a bad night, thanks to sleeping bags pressed against the walls unleashing showers of rime, Wager volunteered to man the other tent alone. There were pros and cons to both arrangements. Wager had the luxury of free movement and the chance to dry out his clothes, but he slept colder than his cozier companions. The first night, as he cooked his dinner, he risked asphyxiation by carbon monoxide (just as Scott and Watkins had on the Southern Journey), alerted to the problem only when he couldn't get his candle to light. After that, all three men cooked dinner in Courtauld and Chapman's tent, where the camaraderie made up for the cramped quarters. On different evenings they read aloud to one another: Palgrave's *The Golden Treasury*, Shakespeare's *King John* and *Troilus and Cressida*, and *Alice in Wonderland*.

On November 13 the men dined on a small piece of seal meat they had brought from base camp, sending Chapman into a soliloquy on deprivation. By then, like their comrades at the Ice Cap Station, the three sledgers were quite sick of pemmican. "It was the most wonderful delicacy I have ever eaten," Chapman raved about the seal meat in his diary. "Most people miss so much in life. You can't realize how marvelous it is to sleep between sheets till you have spent weeks without taking your clothes off in a frozen fur bag on hard snow." Chapman's frostbitten toes now thawed enough to cause intense pain, and he watched in dismay as the nails fell off "and the big toes are raw and stick to my socks." He went on, "I keep stubbing them on the ridges [of snow], causing myself acute agony. We are all too tired to talk."

The daily marches, marginally better after parting from their teammates, were soon compromised by a host of new problems. Besides the difficulty of finding the route-marker flags, the sledges kept overturning as the men hauled them across ridges of sastrugi, and each morning they wasted precious time untangling the dogs' traces with bare fingers that quickly grew numb. On November 15, one of the runners on Chapman's sledge broke in two. "Repaired it with thongs before turning in," he wrote laconically in his diary.

Most aggravating among the men's tribulations, because it was so avoidable, was the discovery on the first night that "we didn't take enough primus prickers from the other people" before the two teams separated. A Primus stove, the choice of expeditions in the cold regions for almost a century, well into the 1970s, came with a crucial tool called the "pricker," without which it often failed to burn properly or light at all. This device was a small metal handle that held a tiny piece of sharp wire, the end of which had to be jiggered up and down in the fuel spout to clear out the inevitable gunk left from each previous burning. Prickers were incredibly easy to misplace, lose, or break. An expedition without prickers might as well have no stoves. The first night, after many other improvisations, Chapman used the second hand from his wristwatch as a substitute. After that, Wager—"a wizard with the primus"— somehow rigged a lasting remedy (we are not told how). Asphyxiation by Primus fumes, however, continued to pose a nightly threat.

Alarmed by the balky progress, on November 15 Courtauld once more pushed the question of his solo occupation of the Ice Cap Station. At the time, the men were camped next to Flag 90. Only at Flag 262 would they arrive at the station. Twenty-one days out of base camp, the men had covered only a third of the distance to their goal. The numbers won the argument. As Courtauld wrote that night in his diary, "Decided to lighten [Chapman's] sledge and only take on enough food for one man—self—at Ice Cap. We shall therefore only take four instead of fourteen boxes there."

Wager still wasn't buying it. In his own diary, he argued, "It was August's idea to stay on by himself and he very much wanted to do so. In fact my insistence on staying he has always attempted to dispute, so that there is no question of who stays at the Ice Cap Station. I am most sorry that I am not."

Only years later did a crucial factor in Courtauld's volunteering to man the station solo come to light, via passages in his diary. On the journey out, he had developed an intense dislike for Wager, though he was too polite to voice his feelings. Months alone seemed infinitely preferable to an enforced isolation with a man he couldn't stand.

According to Courtauld's biographer, Nicholas Wollaston (writing in 1980), Courtauld thought Wager "was too fond of doing other people's work or telling them how to do it." In his diary, he characterized his teammate as "fussy, quick-tempered and rude." Courtauld even suspected that Wager harbored a corresponding dislike for him, though nothing in Wager's diary hints at such a feeling.

There was another reason for Courtauld's keenness to man the station solo—one that he guarded even more privately than his annoyance with Wager. He was deeply in love with Mollie Montgomery, whom he had known since they were both teenagers (she was three years younger). By 1928, his passion had burst full-grown, intensified by his fear that she did not reciprocate his feelings. Shortly before the BAARE, though, the two had gotten engaged. He wrote letters to her throughout the Greenland expedition, and addressed her often in his diary.

As he had headed off to Greenland with James Wordie in 1929, a year before the BAARE, he was stricken with remorse and doubt. Why was he leaving? After the ship steamed away from the port in Aberdeen and Mollie waved her farewell, Courtauld sat down and wrote a letter for the pilot to carry ashore. In it he made a solemn pledge: "If only I had the chance of doing something really big, I feel I could do it—for you." Alone on the ice cap, he ruminated often on that pledge. Enduring the winter months solo in one of the coldest, most alien places on earth, carrying out his duties without fail—that would at last amount to "doing something really big." But here was a motivating spur that no one else should know about—except Mollie.

The men sledged doggedly on through the waning days of November. Reading between the lines of Chapman's diary, one senses that he was still not fully committed to Courtauld's solo mission. But by agreeing to it, and thus jettisoning ten heavy ration boxes, he increased the chances that the daily marches could cover more ground. Far more important than keeping up the weather station through the year was the dire necessity of rescuing D'Aeth and Bingham.

Nevertheless, one setback after another hamstrung the struggle. Random excerpts from Chapman's diary between November 15 and 20:

> Our clothes went hard as soon as we got outside. Most fearful drifts everywhere.
>
> Two miles in five days. It looks as if we shall have to go on till the dog food is finished, then kill off the weaker dogs as food for the others, then—if we find the station—collect Bingham and D'Aeth and manhaul back.
>
> The dogs chewed most of the Lapp thong off [the sledge lashing] last night, which I replaced. They have also eaten all the gut off the snowshoes, which are now useless.
>
> Worst day I've ever had. With the sores in my fork [crotch] and frostbitten toes each step is agony. Several times I just couldn't go on and had to sit down for a few minutes.

Courtauld was having his own problems. His "moccasins" had split open and his feet got dangerously cold. Soon he would develop incipient frostbite in his fingers and toes. The layover days, irksome though they were, provided precious respite from the ordeal of sledging. Inside his tent, Chapman read Stevenson's *The Master of Ballantrae* and *Treasure Island*. Inside his, Wager immersed himself in Shakespeare's *Richard II* and some poems by Dante Gabriel Rossetti. Ever the geologist, he also pondered strata and intrusions. After a rare day that he "thoroughly enjoyed," he noted in his diary, "I spent the morning thinking over the tectonics of the dyke and plateau basalt formation about Kangerdlugssuaq [sic]. A scheme beginning with thinning of the sial by tension fits the facts pretty well." In odd moments, Wager seemed disenchanted with the whole journey, complaining on November 19 about "somebody's remark before we left the wireless that this will be an epic journey—a damned silly one I think in lots of ways." He did not elaborate on that sour judgment.

Yet all the toil was slowly paying dividends. On November 20, Courtauld wrote in his diary, "Found to our surprise that we had done fifty-six miles from the Big Flag, leaving sixty-two to go."

After November 20 a new, almost ruthless determination came over the men. Chapman: "It's just amazing what one can do to these dogs under such conditions. One behaves like an animal and hits them anywhere with any weapon. However they seem quite impervious to punishment either from us or from each other." On the twenty-first, "Had to kill one of my dogs which was too weak to pull. We can't afford food for it. Killed it instantaneously with the blunt side of an axe."

On the twenty-fourth, Tiss, one of Courtauld's dogs, gave birth to a puppy. The dog's father licked the snow off it, but "We relentlessly fed it to another team, and the same had to be done with three other puppies which appeared in turn each time we stopped." Despite having just given birth, Tiss pulled her share of the weight between stops. "Poor brute," wrote Chapman that evening, "but what else could we do?"

The wretched huskies were ravenous for food, their half-rations finally taking their toll. On November 27 several of them tore patches out of Wager's tent, even though it was made out of canvas, and broke into a bag that contained only gear, not rations. The next day, "Hynx too weak to pull: I'm afraid he'll die soon." Two days later, Hynx collapsed in the snow. Rather than kill the dog, Chapman recorded, "I let him run loose, but he got behind and hasn't been seen since."

On top of their struggles with dogs that were starting to play out, the men were finding it harder and harder to find the marker flags. The diminishing hours of daylight made searching for them all the more onerous. A characteristic commentary from Chapman: "Go out on skis to search for flags. We can't go on till we find them. Found them ¼ mile to the left." And the next day: "Search for flags in the morning. Found one at last. . . . Went on for two hours by moonlight, luckily finding the flags."

During these late November days, Chapman observed that one mile per hour seemed to be the team's limit. Long days spent totaling only four and a half or six miles seemed major accomplishments, yet since some of that distance was squandered in back-and-forth sweeps in search of flags, not all of it could be counted as progress toward the station. By now, all three sledges were in woeful condition. The men spent the whole day on November 22 jerry-rigging repairs (cutting off handlebars, for instance, to repurpose them as struts).

By now, both Courtauld and Chapman were afflicted by frostbitten toes, Courtauld also with frost-nipped fingers. Wager had somehow avoided frostbite. The diaries of the other two men grew clipped, the sentences telegraphic, but Wager still had the energy to write long paragraphs characterizing each dog in his team, musing on Greenland geology, and planning other trips in the future. On the twenty-seventh, he jotted down an oddly bland summation of their leader's state of mind: "Freddie is I think pretty tired of this trip. He had only just come back from one [the second relief mission, in October]. But so are we all, I fancy."

On November 29, the men sledged into a region on the plateau that was suddenly more conducive to travel. The surface hardened and the sastrugi became less frequent. With the moon waxing beyond first quarter, the team was able to push on after dark and still find the flags. The next day, Chapman exulted, "How incredible, we've done 12½ miles: a record for the trip." On December 1, despite one of the two Primus stoves refusing to work, which delayed their start, the men still gained ten and a half miles. The day after that, twelve and a half again.

By Chapman's reckoning, they should now be in the vicinity of the Ice Cap Station. They had reached Flag 236, and 237, they believed, marked the end of the route. Courtauld's uncharacteristically long diary entry on December 2 vividly captures anticipatory joy turning into inexplicable defeat:

> Only half a mile to go. We could hardly get the dogs to go again but at last we finished the half-mile. But where was the Union Jack? We searched in all directions expecting every minute to see it, and then warmth and dryness and *food*!! It was freezing over sixty degrees [i. e., thirty below zero] and the wind bit through our clothes as if we were naked. We went on a short distance further and searched again, but failed. At last we had to give it up. . . . We slept little that night.

"Terribly disappointed not to find it to-night," wrote Chapman. In the morning, he dashed out without his windproofs, only to return defeated, his ears frostbitten. "It was the worst pain he had ever known, he said," noted Courtauld. "As regards the disappearance of the Station, we could not make out what had happened."

It was only when Wager ventured out later on December 3 and found Flags 238 and 239 farther west that the men began to comprehend their mistake. "According to the book," Courtauld complained, "the station ought to be at Flag 237." The "book" was the log of the route compiled by the party that had established it and erected the Ice Cap Station back in August. Perusing its pages once more, the men found the discrepancy. "On a back page," Courtauld revealed, "were scribbled more flags up to 262." The men quickly calculated: twenty-five more flags, twelve and a half miles still to go. (None of the three disappointed men addressed in their diaries the obvious question: why hadn't Chapman, who had followed the flags on the relief mission in October, known where the numbers ended?)

Too impatient to wait for the next day, the men sledged through the early night. Chapman performed some remarkable navigating by fixing sights on Arcturus and Vega. At Flag 260 they stopped "when our sledge-wheel told us we had gone far enough." The temperature was forty-two below zero but the moon was high ahead of them. Leaving the sledges, they split up, each headed in a different direction. Just before 8:00 p.m., four hours after sunset, Chapman saw the Union Jack. He hiked back to rally his teammates before actually approaching the station. Courtauld was returning to the sledges dispirited from his own fruitless search when Chapman called out to him with the news. "I have never been so suddenly overcome with joy," Courtauld wrote later that night, "or been delivered in such a short moment from the depths of despair."

In the moonlight the trio sledged up to the station, which showed only as "a low mound of snow with its tattered flag." Parking the sledges, they stayed silent as they walked up to the snow wall surrounding the tent, opened the exit door, scuttled along the underground tunnel, and stopped just short of the door in the floor of the tent. Then one of them barked out, "Evening Standard! Evening Standard!" (The *Evening Standard*, which survives today, was one of London's leading newspapers.)

Moments earlier, Bingham and D'Aeth had heard a noise that they thought sounded like a minor earthquake—the sound of the sledges approaching. "When we stood at the end of the tunnel and showed our faces," Chapman wrote, "they were so covered with ice that Bingham and D'Aeth could not recognize us." Courtauld added, in a masterpiece of understatement, "They were naturally jolly pleased to see us." Within minutes, the newcomers were "wolfing down the brew they had made for us, and giving them lots of news from the outside world."

Exhausted and still in pain from their frostbite, Chapman and Courtauld retreated for the night to one of the little igloos built to house the weather gauges, while Wager stayed with Bingham and D'Aeth in the big domed tent. In the morning, all five men shared what Courtauld called "a really late gentlemanly breakfast." Then, that evening, Chapman prepared a feast, which the men declared the "Ice Cap Christmas Dinner" twenty days before Christmas. He even printed out a formal menu:

MENU
GAME SOUP
SARDINES IN OLIVE OIL
PTARMIGAN
PLUM PUDDING
RUM SAUCE (VERY)
ANGELS ON SLEDGES
DESSERT (DATES AND RAISINS)
MINCEMEAT, JAM, HOT GROG, TEA (WITH MILK)
NOTE—NO PEMMICAN

In his diary, Chapman added, "Though I cooked it, the dinner was better than any other dinner has ever been."

The men hoped to get off the next day, December 5, but another storm delayed the departure. When the relief party had arrived late on the fourth, they had a single day's food left for the dogs. Chapman calculated that they now had only enough food for the four returning men for eight days on half-rations. On the fifth he killed one of the dogs to feed the others. The arduous journey from base camp to the Ice Cap Station had taken Chapman, Courtauld, and Wager thirty-nine days. Getting back in eight loomed as an almost impossible challenge.

Early on December 5, Bingham and D'Aeth learned about Courtauld's plan to man the station alone through the coming months. They were horrified, and argued strenuously against the plan. Each man had a tale about a moment toward the end of their occupation when a trivial event had utterly spooked him. Bingham was alone outside once when the Union Jack caught a gust and flapped loudly. "He bolted back into the tent as fast as he could," summarized Chapman. On another occasion D'Aeth, also outside alone, had suddenly caught sight of a small screen erected to protect a gauge. Thinking the apparition was a strange man appearing out of nowhere, he too dashed back inside the domed tent. During nine weeks of monitoring, neither man had dared venture very far from the station, and that only along the route of the flags. Despite their smooth companionship during their exile from the rest of humanity, a kind of paranoia had crept over their spirits. How much worse it would be for one man alone, they argued, through the darker and colder months.

Wager's view was more equivocal. On arriving at the station and discovering that Bingham and D'Aeth had saved more of their rations than expected, he "again half expected to be staying." But there was still not enough food for two men. And Wager admitted in his diary, "I don't think I should care for more than one month here by myself." Another reason for doubting the wisdom of Courtauld's solitary vigil was that by now, his fingers were frostbitten badly enough that he had trouble closing and opening the snaps and buttons on his clothes.

But the man himself was adamant. Courtauld repeated his whimsical claim that he "rather liked being alone." He had his books, his tobacco, and plenty of food for one man. He cited Watkins and Scott on Labrador trappers wintering solo deep in the backcountry without mishap. For all his own misgivings, Chapman wanted to believe in Courtauld's rationalizations. "I must say it would be a thousand pities to abandon the station now," he wrote in his diary, "since it has been established and maintained with so much trouble." Of Courtauld's volunteering for the vigil, he added, "It's a marvellous effort and I hope to God he gets away with it."

The four returnees got off by 10:00 a.m. on December 6. The first day out, they covered thirteen miles.

In the end, it took Chapman, Wager, D'Aeth, and Bingham fourteen days to dash back to base camp. It was a gutsy performance, and they managed not to lose or have to kill any more dogs. They were hoping to pick up a food dump they had left at Flag 56, but somehow they missed both the flag and the dump. By December 14 they had run out of both candles and paraffin, so there was no light to read by and all their breakfasts and dinners had to be eaten cold. Five days later, on the last stretch, as they struggled down Buggery Bank, two of the three sledges fell apart, broken beyond repair.

Gino and several Inuit met the worn-out party at the foot of the glacier. "He was more than relieved to see us all safe," reported Chapman. But earlier that day, out searching in one of the Gypsy Moths for the returning party, on the first day either plane could take off from the finally frozen fjord, Gino saw that there were only four men in the sledging team, which meant that Courtauld must have stayed alone at the station. Gino dropped "luxuries and dog food" to the men on that homeward stretch.

It is striking that none of the three published accounts of the BAARE—those by Chapman, Lindsay, and Scott—expresses much concern about Courtauld's fate 130 miles away and 8,200 feet higher. On Christmas Day, according to Lindsay, after a sumptuous dinner, "We toasted August's health and sent many a friendly thought across the snows to him." Scott acknowledged that, although Courtauld had food that could last, if carefully managed, into the beginning of May, "for every reason he should be relieved as long before that as possible." But in the next sentence he rather cryptically insists, "The weather was still unfit for sledging."

Gino thought he had the answer, and it had to do with the Gypsy Moths. At the very least, he believed, one of them could fly up to the Ice Cap Station and drop food and messages and letters to Courtauld. At best, a plane could land at the station, obviating altogether the need for grueling relief missions by dog sledge, allowing an effortless rotation of monitors through the winter months.

As early as November 11, the base camp fjord had been inundated by a fierce blizzard. Returning from an aborted trip to one of the Inuit villages, Lemon and Watkins got bad news from the locals. "The Eskimos cheerfully told them that the blizzards would last till March," Chapman summarized, "getting worse each time till they would culminate in a really prodigious hurricane at the beginning of March."

The natives knew what they were talking about. Beginning in December, storms succeeded one another almost uninterrupted, and even on clear days the wind tore violently across the fjord. It would not be until February that a Gypsy Moth was able to make the first of several attempts to fly to the Ice Cap Station. None of those flights would come off as planned.

*   *   *

On the morning of December 6, Courtauld had watched his four teammates start the long journey back to base. "Coming out again an hour later," he wrote in his diary, "I could just see them as a speck in the distance. Now I am quite alone. Not a dog or even a mosquito for company."

At once Courtauld set about drying his clothes and sleeping bag, since everything was "full of ice." During the first few days, the routine of recording the weather gauges every three hours from 7:00 a.m. till 10:00 p.m. kept him occupied, and his mood was upbeat. But on December 8, only the third day of his sentence, for the first time the sun failed to clear the southern horizon—"and I suppose will not until the middle of next month."

The next day, Courtauld made an unpleasant discovery. Bothered during his sleep by persistent itching, he changed his underwear, only to discover that they were crawling with lice. "This is what comes of lending one's sleeping-bag to Eskimos," he jotted in his diary. Disgusted, he laid out the underwear in the snow outside the tent "in the pious hope the cold will kill them." At first, the remedy seemed to do the trick; but the lice would return as the larvae deposited all through the inside of his sleeping bag hatched.

Courtauld kept his spirit buoyed by relentlessly "tidying up." The diary records his daily chores and small delights. On December 11, in the first of many such annotations, he wrote, "Reading [John Galsworthy's] *Forsyte Saga*, Vol. II—V. G. [very good]. Even better than Vol. I."

And: "Opened pea-flour and marge today. Found jam made out of cocoa V. G., much better than drinking it."

But he also reported "Toes hurting, also fingers." And already he was finding the cold inside the domed tent unpleasant. With the Primus firing, the inside temperature rose to sixty degrees Fahrenheit, but with the stove off, it settled around thirty-five degrees. Outside the shelter Courtauld was recording temperatures as low as fifty-six degrees below zero. The only direct link between the double-layered interior and the outside was a metal tube two inches in diameter that poked like a periscope through the roof. It had been devised as an emergency source of air in case the tent got completely buried by snowdrifts. On December 12 Courtauld stopped up the tube, hoping to make the tent a little bit warmer.

That same day, for the first time he noted a problem that would come to be the bane of his existence: "Entrance to tunnel blocked up when I went out for 10 p.m. obs. Had to dig my way out." He added, "This weather won't help the others getting back." At that moment, Chapman and his three companions had covered more than seventy miles along the route home, and they were looking forward hungrily to the food dump at Flag 56 that they were destined not to find.

Smoking his pipe (a gift from Mollie Montgomerie) was a keen daily pleasure, though already he was worried about when he might run out of tobacco. "Found I am only smoking 1.7 ounces a week," he wrote on December 14. "Tobacco should last at this rate seventeen weeks." The other great pleasure was reading. The same day, he noted, "Reading [Stevenson's] *Black Arrow*, [Stefansson's] *Friendly Arctic*, Isaak [Walton]. All V. G. What I shall do when I have finished all the books God knows."

On that day (a Sunday), his ninth alone, for the first time he recorded the kind of prolonged pipe dream that would serve as temporary escape from the monotonous extreme of the ice cap. "Made out a list for a chap's dinner when I get home," he reported. (Such flights of fancy are par for the course among expeditioners in hostile environments, and so often they revolve around food.)

That fantasy was so vivid that Courtauld drew a table plan for the dinner in his diary, with places for thirty-six friends (all men). He lovingly detailed the menu, ranging from oysters as an hors d'oeuvre to courses of grouse and tournedos of beef to pancakes flambés for dessert. He even specified the wines for each course and the brandy to be served as digestif.

Early on, he tried to calculate how long his food supply might last. In theory, if he cut his daily intake to a bare minimum, he should be able to eke out his rations until the beginning of May. But Courtauld rebelled against such self-discipline. Arbitrarily he set March 15 as the date on which, at the latest, a relief party would arrive. As he eventually wrote, "I prefer to eat my cake rather than have it. *Carpe diem* was a tag which served as an excuse whenever I was hungry."

On December 13, he played a game of chess against himself. (The diary does not record who won.) He also rebandaged his toes ("both seem quite dead but are gooing"). A few days later: "Tonight unbandaged toes. Unpleasant sight. Left toe-nail came off. Other will soon, I expect."

Already on the thirteenth, he recorded further troubles with the exit tunnel: "Entrance to tunnel completely stopped up and had to dig myself out for every observation." The complaints multiplied during the following days. On December 16, "Had a job to get out of the house this morning. Found I was digging up into a vast snow drift, so had to make a hole vertically upwards, and after some time burrowing managed to scramble out."

Still, through most of December, Courtauld's morale stayed solid, as his complaints about his toes and the exit tunnel were balanced against moments of happiness or awe ("Aurora wonderful tonight, like purple smoke wreaths twisting and writhing all over the sky"). But December 21, the shortest day of the year, triggered a kind of existential revelation of his self-imposed predicament. "At ten o'clock [p.m.] it was completely still," he wrote. "The silence was almost terrible. Nothing to hear but one's heart beating and the blood ticking in one's veins."

Christmas Eve brought with it the first wave of melancholy. "How wonderful it would be at home. . . . This time last year was such a marvelous Xmas." That day he made a grim discovery. Somehow two of the big cans of paraffin had sprung leaks, and four gallons of the precious stuff had drained away. The paraffin fueled both the Primus stove and Courtauld's lamp, which gave off some heat of its own. Until December 21 he had often run the stove just to heat up the tent. Now he would have to restrict the Primus to cooking only. And at first, he dared not calculate when he might run out of paraffin altogether.

Christmas Day brought an even stronger onset of melancholy. "I wonder what they are doing at home," he wrote. "I really do not miss the good things of Xmas very much. Though I would rather like a bit of fresh meat and a mince pie, and even more a bit of plum pudd."

Even more than Christmas at home, he missed Mollie. So private were Courtauld's feelings about his fiancée that in his diary he gave her the single-letter pseudonym of "W." "B[etty, his sister] I suppose in Switz. I wonder if she has taken W. with her." "Pipe (W.'s) going V. G." And on December 31, among his five resolutions for the New Year: "Get home and ask W. to marry me." (This despite the fact that they were already engaged.)

In another extended fantasy on the day after Christmas, Courtauld spelled out the specifications for a settled life after he and Mollie had married.

I think I should like a small house in Suffolk between twenty and sixty miles from H[ome]. Should be near for trains and near the sea, preferably Pin Mill. No land except a garden and fewest possible servants. No waiting at table. If any money would rather spend it on a boat than a house. . . . A Brixham trawler would be almost ideal but would probably cost a lot to make it habitable. It would be better to keep her at Falmouth or somewhere on the south coast. . . .

It sounds pretty swank and genteel, but compared to the surroundings in which Courtauld had grown up, that fantasy life figured as almost bohemian.

On January 1 he made a list "of books worth reading taken from Ice Cap library"—implying there were even more at hand he thought not worth reading. The list is twenty-seven titles long, and does not include some of the ones (*The Forsyte Saga* or *Black Arrow* among them) that he had already rated V. G. It's an eclectic roster, ranging from the seriously scholarly (the four-volume *Cambridge History of Empire*) to novels by Fenimore Cooper and Dickens; from adventure stories such as Dana's *Two Years before the Mast* to *Under Sail*, a lively account of a sixteen-year-old on a sea voyage from New York to Hawaii; as well as popular astronomy books by Sir Arthur Eddington and Sir James Jeans, along with a smattering of poetry anthologies.

Yet in Courtauld's list of New Year's resolutions, along with asking Mollie to marry him, were "(4) Give up exploring" and "(5) Collect a library and study: (a) English literature and poetry; (b) Music; (c) Polar exploration with a view possibly to try to write a book about it."

Those jottings may have been partly tongue-in-cheek, but after almost a month alone in the cold and darkness, despite books and chores and pipe dreams, Courtauld may have been starting to feel the oppressive toll of his solitude. On December 27, a strange event occurred that deeply disturbed him. He described it in his diary:

Just getting to sleep again after 7 a.m. obs. when there was a soft rumbling close to my head which increased and ended in a dull crash. It flashed across my mind as it began that the weight of the snow was too much and the whole house was going to come in on me. However nothing happened so I concluded that the tunnel had fallen in and that I should have a job to get out as the spade would be buried. However that was not so. . . . Hope nothing further happens tonight.

In the absence of a plausible explanation for the rumbling sound and the crash, it is not surprising that the two catastrophes that must have been lurking in his head came first to mind: the collapse of the tent and the sealing shut of the tunnel. But when neither proved to be the case, his fears migrated into the realm of the uncanny and the unknowable. Of such uncertainties, paranoia is born. (Later commentators have speculated that the disturbance might have been caused by the sudden settling of massive layers of snow and ice around the tent— almost like a horizonal avalanche rippling unseen beneath the ice cap's surface.)

But on January 4, one of those catastrophes seemed to become reality. Courtauld woke on another Sunday to a furious gale shaking the tent. He crawled through the trap door and found, as he had expected, that "the tunnel was snowed up." He dug it out at 7:00 a.m., and again at 11:00, 1:00 p.m., and 2:30. On his last excavating mission, he discovered that "the back of the tunnel was so full of snow caused by digging out the entrance that I could scarcely wriggle up to it. . . . At 3:30 I found that I could no longer get the snow back from the entrance." He was effectively trapped inside the tent.

Whether Courtauld felt an onrush of panic that afternoon or not, he forced himself to write a few more rational lines in his diary. "If I cannot get out, I shall have to stay in and the met[eorology] will have to go hang until the wind drops. Then I shall have to find some way of getting out." But now, the other two fears surged to the forefront of his thoughts—the chance that the metal air tube in the ceiling might get plugged or that the roof might collapse.

If either happened, well, then. . . . Striving for jaunty stoicism, he wrote, "My end should be peaceful enough, and I have four slabs of chocolate to eat during it. Anyhow it won't be attended by the fuss and frills one's pegging out at home would."

NINE

Winter with the Inuit

IF THE BAARE men gathered at base camp after December 19 seemed somewhat blasé about Courtauld's safety and well-being in his solitary outpost far up on the ice cap, the tone was set by Watkins. As Jamie Scott later wrote, "Gino showed no anxiety. He knew that most of the dangers of solitude are engendered in a man's imagination, and he knew that Courtauld's mind would be too full of his work and hobbies for anxious thoughts to have a breeding-ground. Even had Gino doubted the wisdom of this decision his lips would have been sealed, for he had given [Chapman's] party a free hand to do what they thought best. In any case there was nothing to do about it till the spring."

Nothing to do about it? Perhaps because the relief mission that had left Courtauld at the station had had such a desperate journey, and perhaps also because the men trusted the Inuit forecast of bad weather throughout the winter, they all seem to have agreed that (again in Scott's words) "there would be . . . no need to send up another relief party until March or April, when the storms should have subsided."

Given that Watkins was the leader of a complex fourteen-man expedition, it's surprising how peripheral a role he had played between July and December in the main projects of the BAARE. On the Northern Journey, while his teammates explored Kangerlussuaq Fjord and the coast south of it, Gino had spent much of the time in the Gypsy Moth, reconnoitering the coastal mountains and capturing stunning aerial photos of undiscovered places. Although he had carefully designed the double-layered domed tent that would constitute the Ice Cap Station, and specified the weather gauges that the monitors should consult daily, he had participated in only one of the three journeys up to it—the second relief mission that swapped out Riley and Lindsay for Bingham and D'Aeth. And he went along on that trek mainly to get to an ideal point from which to launch the Southern Journey.

That two-man triangular ramble across the ice cap had been Watkins's chief exploit so far on the expedition. It was also the most disappointing, after storms and failing dogs had forced him and Scott to cut it short by more than half. And Scott had admitted that the scientific results of that sledge trip to nowhere had been negligible.

Gino had deliberately brought to Greenland far less in the way of rations than the men would need to sustain themselves for a year. Living off the land had been one of his cherished ambitions. What he hadn't counted on, though, was the bad luck that Inuit seal-hunting in the autumn months had provided only enough meat to feed their own families—as well as the inevitable consequences of turning the base camp hut into open house for all visitors. The Inuit naturally expected the same hospitality from the white men as they offered in turn in their stone houses. And British food was a novelty and a luxury to palates used to seal, seal, and more seal, with a few fish or a ptarmigan thrown in now and then.

As Scott wrote about the installation of Arpika, Gitrude, and Tina in the hut attic, "Our young staff worked for love and for the right to play the gramophone, on and on, grinding their favorite records until there was little left but noise. Very often their friends and relations came to visit them, ate with us, laughing at the forks and spoons, and slept on the floor in rows." The BAARE men also threw parties for teammates' birthdays and for Christmas, with Inuit usually in attendance. "The white men shared our small supply of alcohol," Scott commented, "but remembering Labrador, we gave none to the natives." (In the coastal ports of Labrador, Scott and Watkins had witnessed the impact commercial fishing boats had had on native culture, ranging from booze to smallpox and syphilis. What Scott doesn't mention is that in Greenland, the Danish government strictly forbade giving alcohol to the Inuit.)

An intriguing note regarding the gramophone: All the records the BAARE men brought to Greenland blared out jazz, show tunes, and popular songs. Six years later, living with his Inuit "family" in Kangerdlugssuatsiaq Fjord, Paul-Émile Victor noted the same addiction among the Inuit to the gramophone, but added, "A tremendous success with them was jazz, which made them laugh. But their real preference, when I was there and allowed them to use them, was for the records of Beethoven, Chopin, or Bach."

Fortunately, before the expedition had even started, Gino had received a telegram from Scott, who was buying the sledge dogs on the west coast, lamenting the fact that he hadn't been able to purchase as much dog food as he'd hoped. In Copenhagen, as a fallback reserve, Gino had bought several hundred cans of boiled beef. As Chapman would later write, "This proved an absolute godsend, and we lived on the boiled beef half the year at the Base. . . . Luckily, the Eskimos liked it even better than seal-meat."

The best seal-hunting would come only after the fjord froze over. But maddeningly, through the end of November and the first half of December, temperatures stayed unusually warm. The fjord would start to freeze, only to have the next storm blow all the new ice out of the inlet. The full service of the Gypsy Moths also depended on a frozen fjord, on which the planes could take off and land with their ski undercarriages, without which landings on the ice cap would be impossible.

In November and December, then, there was a lot of down time around base camp. Nobody was really idle, for there were always chores to do, such as chopping the six-inch coatings of ice congealed by sea spray off the crates stored outside the hut, or building new toboggan-like sledges with light runners and minimal friction for coastal transport. Watkins also directed the construction of a hangar to protect the Gypsy Moths—no mean architectural challenge, at twenty-five feet square with a height of eighteen feet beneath the solid roof.

Christmas called for a gala celebration. Breakfast was tomatoes and sausages, and lunch was skipped to save room for a dinner of successive courses of rabbit soup, salmon, beef tongue, asparagus, *petits poussins* ("little chicks," presumably ptarmigan or grouse), and Christmas pudding. After dinner the men opened a special package from their mentor, Frank Debenham, that turned out to contain letters from home, new gramophone records, a selection of novels, and best wishes from the BBC and the Prince of Wales. With the opening of Debenham's Christmas gift, the party had only begun, for dancing (presumably with Inuit partners, among them certainly the three young "maids") went on until 3:00 a.m.

Chapman marveled at how well the men got along, even on slow days in the overcrowded hut.

> It was amazing how little we quarreled. . . . We had representatives from England, Scotland, Ireland and Australia—not forgetting the island of Jersey. The variety of our religious opinions included complete Agnosticism, Nonconformity and Anglo-Catholicism. We did not appear to have much in common; why were we there? . . . For some it was a career; for some a holiday, and for others, the realization of an ideal.

Compared to nearly all the men in charge of polar expeditions before him, Gino practiced a style of leadership that was radically egalitarian. Of course, as the youngest member of a fourteen-man team, he could hardly adopt the rigid authoritarian posture of a Robert Falcon Scott or a Robert Peary. During the many months in Greenland, some of his teammates occasionally grumbled at Gino's laissez-faire approach. Even Courtauld, the most equanimous of men, let slip a parenthetical gripe in his diary during his first month alone at the Ice Cap Station, as he mused that "it seems that on this show no prearranged plan ever gets fulfilled." Martin Lindsay, in particular, later criticized Gino for failing to include a truly expert surveyor/navigator in the team, and for sending out the last two relief parties onto the ice cap without clear instructions. But as a trained army officer, Lindsay swore by discipline as the only way to get things done.

All in all, the unflinching loyalty of his teammates through fourteen months of ambitious journeys and near disasters bespeaks a triumph of leadership as blithe as it was enigmatic. In his 1935 biography of Gino, Jamie Scott paused in the telling of the BAARE saga to devote a whole chapter to the puzzle of Watkins's success as commander of "the greatest British expedition to the Arctic for half a century."

Scott quotes Sir William Goodenough, president of the Royal Geographical Society, offering this valedictory praise of Watkins: "Leadership came to him naturally. It is a common saying that in order to command one must first learn to obey, but here was one who appeared to leap fully equipped from the levels of boyhood to the eminence of a man's directive power."

Scott agrees with Goodenough, yet recognizes that that formulation fails to explain the secret of Gino's simultaneously filling the roles of apprentice and leader. For Scott, Gino's gift was that he was always intensely receptive to the counsel of more experienced veterans. As a rock climber, Gino always led the hard pitches, but "in gatherings of mountaineers, he listened instead of talking so that he quickly learned all that he should know." On Edgeøya, despite knowing almost nothing about the landscape beforehand, Gino "was successful because he could sum up positions quickly and act without hesitation." In Labrador, as Scott saw firsthand, "Everyone [Gino] came in contact with was gratified by his respectful interest in all they said and, without fully realising it, they did what he wanted them to do and taught him all the useful knowledge they possessed."

"They did what he wanted them to do." As noted previously, Gino recruited Scott for the nasty job of buying sledge dogs in West Greenland and then kenneling them on a barge off the coast of the Faroe Islands by slyly musing, "I wonder if we could find a man who knows as much about dogs as you do?"—and won Scott's grudging admiration for that sleight-of-tongue. Likewise Lindsay agreeing to reinforce the ill-starred third relief mission after Gino offhandedly remarked, "I say, Martin, do you mind going up with Jamie to relieve August?" (He added: "Are your moccasins in good order, because it'll be hellish cold up there.") Rather than bridle against an order he was loath to obey, Lindsay admired how Watkins "chose his man for every job of work and then asked him to do it in such a way as to inspire him to a relentless devotion to duty."

In the loosely egalitarian milieu of the BAARE, it was thus not surprising that while most of the men addressed one another by last name, Watkins was "Gino" to everyone. "That alone cramped the possibilities of service discipline," Scott wrote, "for one cannot call a man Sir and Gino in the same breath without laughing." Scott remembered that in Labrador a man had once called his partner "Boss," with the result that "Gino had been a little embarrassed and very much amused. That was not his name, so why address him so?"

The key to Gino's leadership, Scott believed, was his sense of humor. In the early days of the expedition, when the men were just getting to know Watkins, it would have been natural, Scott thought, for his teammates "to treat him with at least the outward deference they were accustomed to show to a commander." But Gino wanted no part of it. "He took trouble to climb down from this uncomfortable eminence," Scott wrote, "by telling stories against himself, flirting with the Eskimos, posing as an utter Philistine or joining in every menial task."

Another key to Watkins's leadership style was that even in the grimmest predicaments, he stayed eerily calm. Panic was not a state of mind with which he was acquainted. Scott insists that only once during the BAARE did Gino get really angry at a teammate. One would like to know who that antagonist was, and what the conflict was about, but the gentleman's code of the day prevailed, as Scott discreetly summarized, "Once only Gino remarked that a man was beginning to behave badly and that he thought he would have to have a row with him. He had his row, in a roaring temper, so it seemed, and afterwards they were far stronger friends than they had been before."

Gino's casual air about logistics or decision-making was belied, in Scott's assessment, by the imaginative care he had put into every aspect of the team's planning. "The clothing was light and warm," Scott commented, "while the sledging rations—Gino's most striking innovation— were excellent; no one had been really hungry on any journey and there had been only one case of serious frostbite." The tents, designed by Gino, were lighter and sturdier than those that any of the men had camped in before. The double-layered dome tent for the Ice Cap Station, modeled somewhat after Inuit igloos and snow houses, seemed ideal, until its fatal flaw of the trapdoor exit started to work its harm. Even the base camp hut, with central kitchen and double walls, stayed warm and comfortable through the winter months.

Another aspect of Watkins's leadership style had to do with what Scott called Gino's "cold practicality." As an example, Scott cited a low moment on the Hopedale traverse in Labrador when, running out of food, Gino had mentioned that if he died he fully expected to be eaten by his surviving teammates. "When I demurred, he added, 'Well, I'd eat you, but then, of course, you are much more fat and appetising.'"

Scott would remain a lifelong admirer of his erstwhile best friend. In the "Leadership" chapter of the biography, he boiled down Gino's style to two principles. "He was a young man who set out to enjoy himself," he wrote, "and to make others enjoy themselves as well, because he believed that people worked better when they were happy." And:

"Briefly, his method of leadership was to train each man to be a leader: his ideal exploring party consisted of nothing else."

Yet Scott sensed that these formulas failed to capture the essence of a *modus explorandi* that was, at its core, a paradox. He ends his disquisition with a conundrum: "He followed the path which he had chosen, enjoying every step, quick to shock, slow to offend, . . . leading without looking back because he knew that we would follow him. Both as a friend and a leader he had always something in reserve, some depth which gave occasional proof of its existence, but which even he did not understand."

* * *

More than any of the other men on the BAARE, Gino wholeheartedly embraced the idea of learning from the Inuit how to adapt to the extreme environment of Greenland. In Scott's summation,

> After Christmas he went to live for a few weeks at the southern settlement and learned there how to hunt seals with a long harpoon, in open leads and at the breathing-holes. He discovered that he was at least as patient as his teachers and a better shot than most of them. He ate the food they gave him, from the raw blubber to the boiled seal guts, as readily as he adapted himself to all their customs. Every night he added words to his vocabulary and learned something more about an intelligent but superstitious people who were so well adapted to the conditions under which they lived that they were seldom uncomfortable and always cheerful.

(One suspects that veteran hunters such as Nicodemudgy might have raised an eyebrow about Scott's claims for Gino's patience and marksmanship.)

Yet even as he went to school with the Inuit, Gino experimented with ways to improve his men's performance. In early winter, he tried out a new method of hitching dogs to sledges, rejecting the Inuit "fan-style" configuration (several separate lines each attached to two or three dogs) in favor of a "single-trace" pattern, with all the huskies' leads attached at regular intervals to one main line. Gino quickly concluded that his own arrangement would be more efficient, especially on the ice cap, even though, according to Scott, "A Danish governor who had been many years in Greenland had told him that it would be impossible to train the dogs to do this."

Gino was not the only team member to try to learn to hunt seals Inuit-fashion, but few of the BAARE men could come close to the technique of even the least-accomplished natives. Their efforts provided much entertainment for the locals. Lindsay's vignette captures the spectacle:

> On the rare occasions that we got [a seal], we sometimes used to think it necessary to complete the illusion that we were hunters—by flenching and cutting up the kill ourselves. We waded unswervingly through blood, getting more and more mixed up among the organs, to the great delight of the girls. They themselves could do it in five minutes, but it was only after much practice that we managed to complete the operation in three times as long. Kneeling to cut up a seal makes one very tired unless one is something of a supplicant.

Even as they grew somewhat competent at hunting seals with harpoon or rifle through breathing holes, the BAARE men were limited to pursuing their prey on "the old ice at the top of the fjords." They often watched seals cavorting in the open waters offshore, tantalizingly out of reach. Over the centuries, the Inuit had perfected the art of hunting seals from kayaks—arguably the most brilliant, and at the same time the most perilous, of all the natives' adaptations to the stern conditions of the Greenland coasts. With his enthusiasm to learn from the indigenes, Gino became almost obsessed with mastering that skill himself, and several of his more adventurous colleagues went along with the experiment. But hunting from kayaks would have to wait until spring.

Meanwhile, spurred on by the need to feed their own crew through the winter, the seal hunters on the team ventured more and more daringly across the landfast ice. This was trickier and riskier than it looked. As Lindsay wrote,

Sea ice is nasty, treacherous stuff, always to be treated with respect. It has none of the integrity of freshwater ice, which will crackle and bend before it gives under your weight. You may step on sea ice nearly three times as thick as the bearing surface of freshwater ice and go straight through, just as if it were toffee.

To deal with these uncertainties, an Inuit man carried an ice-spear, a thin, six-foot-long wooden pole with a metal spike at the end, with which he probed the surface ahead of his cautious steps. But the BAARE men made the mistake of thinking that a ski pole or an ice axe would serve just as well. The most treacherous surface was new ice covered with a thin layer of snow, for there "you cannot see what you are walking on." And, as Lindsay gloomily warned, "Your first mistake is likely to be your last."

Percy Lemon was the first to blunder. Out on the sea ice with Mikadi, a young Inuit who was perhaps a little too casual himself, he followed the native across a dubious patch of mottled snow ice. Suddenly both men broke through—but both quickly recovered by getting their elbows onto harder ice and wrenching their bodies out of the water. At this point Mikadi, quite chastened, ran back along their outward trail; but Lemon, emboldened by his easy escape, with only a ski pole for probe, pushed on. The ice ahead, he later said, "looked the same."

This time when it broke beneath him, he found himself neck-deep in a broad pool of thirty-two-degree water. He swam five yards, got his forearms onto stable ice, but couldn't lurch free. So he yelled for help. According to Lindsay, "Mikadi ran back to his sledge for some line, and then, jumping cleverly from floe to floe, got across to Lemon and pulled him out." Lemon admitted he couldn't have held on much longer.

On expeditions, the near-tragedies of one day turn into the camp jests of the next. Lindsay wrote up a mock epitaph:

CAPTAIN P. LEMON
STEPPED ON A PIECE OF ICE TO SEE IF IT WOULD BEAR.
IT DID NOT.
AGED 32.

The quest for seals, however, was so crucial for winter sustenance that despite Lemon's close call, other men kept treading across the coastal floes. "Most of us at one time or another went through the ice," Lindsay confessed. But in *Those Greenland Days*, he dropped the cavalier tone to give a sober report on the terrors of plunging through the unstable surface:

When we were alone and had to get out by our own unaided exertions, we experienced what it is like to be thoroughly frightened. The water is so cold that it seems to paralyse all your muscles, and whether you can swim or are anything of an acrobat is almost immaterial. It is a horrible thing to be afraid—to be shaken by a spasm of fear which comes so suddenly that the controlling forces of the brain have not time to attune themselves. In one instant a man seems to become nothing more than a quivering animal.

The utter dependence of the Tunumiut on the seal was the greatest threat to the people's survival. Every prolonged famine, and the mass starvation that it triggered, was caused by a season or a year in which seals failed to show up in the fjords where the Inuit had always hunted them. Even the savviest *angakut* seemed powerless to predict or prevent those catastrophes.

But the dozens of ways the Inuit used every part of the seal bespoke an adaptation as ingenious as it was fragile. The BAARE men were awed by what they saw. Not only did the people eat every part of the animal, including brains and intestines, but they found myriad uses for the skins. The bearded seal (*Phoca barbata*) supplied the coverings for umiaks and tents. Kayaks, however, were best constructed out of the skins of smaller seals such as the bladder-nosed (*Cystophora cristata*) or the so-called Greenland seal (*Phoca groenlandica*). Those animals' skins were also turned into anoraks, waterproof trousers, and boot soles. The skins of the smallest seals, known as "fjord seals," could be turned into dog harnesses, boots, and gloves. Chapman admiringly described another piece of seal-craft: "The sinews are carefully cut out and separated, then plaited together, being rubbed up and down the cheek to twist them together. As seal sinew stretches when wet, it makes the only known perfect water-tight stitching."

Blubber was both eaten raw and used as fuel for stone lamps. The oil served to pickle berries. And every drop of seal blood was poured into the stomach skin to make blood sausages. The BAARE men ran the spectrum of squeamishness, with Gino on the high end of adaptability, but Chapman spoke for most of them when he observed that "when boiled, the intestines themselves are quite good, but, as far as we were concerned, only if there was nothing else to eat."

As winter deepened into January, Gino had big plans for the Gypsy Moths. But on January 10, he got terrible news. Just before New Year's Day, Iliffe Cozens had flown the Angmagssalik radio operator to the base camp fjord, as a thank-you for the many services that man had rendered the team by passing on messages from England and Denmark. A day or two later, after a big party in the hut, Cozens had returned the man to the village. But as Cozens lingered in Angmagssalik, a ferocious three-day storm descended on the east coast. On January 6, the plane broke loose from its moorings, careened across the water, and smashed into broken floes of solid ice. In Cozens's view, the plane was smashed beyond repair, though the engine was intact. To add to the pilot's agony, because Lemon was off on a visit to an Inuit settlement, four days passed before he could radio the bad news to Gino.

On January 12, Watkins and Hampton flew the other plane to Angmagssalik to assess the damage. It lived up to their fears. In Chapman's summary, "One wing was reduced to matchwood, while the other was considerably damaged, the tail had been carried away and the whole longiron was bent." But Hampton, a jack of all fix-it trades, had an arsenal of spare parts, and thought he might be able to get the plane flying again in six weeks or so. Gino was dubious.

This mishap put the grand hopscotch flight across Greenland and the Canadian Arctic to Winnipeg, planned for February or early March, on indefinite hold, and maybe in the trash bin of fond hopes. Depots of fuel and oil had already been cached at key points along the route. But it would be far too risky to attempt the daring flight without a backup plane, and besides, one of the Gypsy Moths was essential for Gino's plans to resupply the Ice Cap Station and maybe to relieve and exchange its resident monitors.

With the storms continuing one after another, just as the Inuit had predicted, the team had to wait until February 8 to make the first flight along the flagged sledging route toward the Ice Cap Station. From the ice in the base camp fjord, D'Aeth and Chapman got airborne. Their plan was to fly to the station and drop "letters, books and luxuries" to Courtauld. Chapman was dazzled by his first view from the air of terrain he had struggled across on foot for half a year. "Stumbling about among the floes," he wrote, "I had seen through a glass darkly, but up here all its secrets were revealed."

The plane circled, gaining height, until at 9,000 feet D'Aeth headed inland. Chapman's rapture intensified: "I could see the coastal mountains all spread out below like a dream country, and far away to the

North the range of [Mont] Forel, jagged and forbidding." But a little more than halfway to the Ice Cap Station, the plane ran into a solid bank of ground-hugging cloud. Too heavily loaded to fly above it, the Gypsy Moth had to turn back. On the way home, the men spotted Lindsay and Lemon, who had sledged all the way up to Flag 56 in hopes of retrieving the wireless set cached there by the relief party in November. They not only failed to find the radio—they couldn't even locate the flag or the big depot piled up beneath it.

The next day, back in the hangar in the base camp fjord, D'Aeth and Cozens were searching for an oil leak in the plane when they forgot that they had left the ignition on. D'Aeth started to turn the propeller with his hand, only to have the engine fire. The propeller, whirling full speed as it would before taxiing or takeoff, caught D'Aeth on the elbow and knocked him across the hangar. Several men carried him up to the base camp hut. Hampton, the team doctor, later discovered that the elbow was broken, though it eventually healed well without being set. But D'Aeth was hors de combat as a pilot for several months, though he was lucky not to have been more severely injured. As Chapman commented, "Not many people are hit by propellers and live to tell the tale."

Storms swept the coast with monotonous regularity for more than two weeks after D'Aeth and Chapman's futile attempt. It was not until February 25 that another flight set out to reach the Ice Cap Station. This time Cozens piloted the Gypsy Moth, with Scott as navigator. An hour and twenty minutes later, having covered what they calculated as the 130 miles to the station, they could spot no sign of it. It was only after they completed the return journey in one-third the time of the outward push that they realized they must have been bucking a fierce head-wind as they had headed west, and had thus come nowhere near the station.

The next day Cozens made a routine hop over to Angmagssalik to deliver the colony's pastor, who had made a short visit to base camp. As he landed, he smashed into a hidden ice hummock. No one was hurt, but the undercarriage was driven straight through the fuselage. The weary Hampton, slaving away nearby on the first wrecked plane, sighed as he surveyed the latest smashup. He thought he could fix it, but it would take four weeks. And he still had three weeks of work to get the first plane even close to airworthy.

With the second crash, the grand dream of the flight to Canada had to be scrapped for good—"a great disappointment to all concerned," as Chapman blandly put it.

Now it became imperative that a sledging team be mobilized to go to the relief of Courtauld. By February 25, the man had been alone in the domed tent for eighty-three days. It was high time his teammates went to his aid, and maybe to his rescue. And though no one said so out loud, some of them must have wondered whether it was already too late.

*   *   *

On January 5, after discovering that he could not unblock the snow that had piled up in the exit tunnel, trapping him inside the domed tent, August Courtauld tried a desperate alternative. He burrowed partway down the tunnel, then turned right and entered one of the two side tunnels that previous monitors had constructed to provide shortcuts to the gauges. It too was blocked with piled-up snow, so Courtauld dug straight upward. After a huge effort, worried that the snow that had drifted over the whole camp might cave in and smother him, he broke through, saw the sky, and forced his way out. "THANK GOD," he wrote in all caps that evening in his diary.

The brief imprisonment had rattled him. The next evening, about 11:00 p.m., in total darkness, half asleep, he suddenly heard Mollie calling. "Aug, Aug," she seemed to moan. To counter the nightmare, he immediately jotted down a detailed plan for a yachting cruise with Mollie down the west coast of England for the next summer.

Through the next week, Courtauld kept his jottings brief and prag-

matic. He read Sir Walter Scott's *Guy Mannering*, which he rated
"V. V. G." But "descriptions of food make me writhe, worse than *Forsyte Saga.* . . . The Potage à la Meg Merrilees sounds marvellous." On January 14, however, he was spooked by a day of high winds that suddenly dropped to dead calm. "Now still as death and dark as pitch," he wrote. "Barometer dropping like a stone. Suppose something pretty unpleasant is about to happen."

With almost no emotion, he noted in his diary the return of the sun above the horizon on January 12. On the sixteenth, as he was outside tending to the gauges, he was stricken with faintness. For a short while he feared he wouldn't be able to get back *inside* the tent. When he did, he allowed himself "a short rest of six hours or so," while his pounding heartbeat slowly returned to normal. But he felt guilty for missing one of the observation runs. His anxiety now focused on his fear of "house showing signs of collapsing. Wish I could get snow dug off the roof. Would have done so yesterday if I had not felt so bad."

That evening, "Finished *Guy Mannering*. Jolly good book." A few days later, "Just finished *Jane Eyre*. It's a great book, one of the best I have ever read. How she can have written it in those early Victorian days is past imagination."

On January 23, he noted for the first time a worry that would gradually intensify in the days to come: "Only ten gallons of paraffin left now." He had already cut down the time he allowed his lamp to burn, and since December 21, when he discovered the loss of four gallons of fuel due to leakage, he had stopped firing up the Primus just for warmth. This meant, of course, all the more time confined to his sleeping bag, and countless hours in the long night when it was too dark to read.

But reading continued to be his greatest pleasure. On January 28,

Finished *Wuthering Heights*. Read it before in Greenland 1926. It is a fine book but I don't like it so much as *J. E.* although, I suppose, a greater book. The plot seems too unreal, the tone too dreary and the characters too monstrous. Now reading Pepys and *Vanity Fair*. The latter, having read it before, seems excellent.

"Pepys" was Samuel Pepys's diary of life in London from 1660 to 1669, a plain but vivid account of doings in one of the world's greatest cities, encompassing the terrible months when bubonic plague killed one quarter of the metropolis's 400,000 citizens.

That same entry records two new sources of anxiety. "Through my carelessness," as he castigated himself, he had broken both the station's barometer and its maximum-minimum thermometer, the former by accidentally dropping a cooking pot on it, the second by merely trying to clear the instrument of snow. The second concern sprang from deep in his psyche. January 28 marked Courtauld's fifty-fourth day alone. Up to that date, if he had started to be assailed by fears of abandonment, he had kept those thoughts inside his head. He had already hung on far longer than either Riley and Lindsay or D'Aeth and Bingham before wondering in his diary what would happen if no relief party came— despite the immeasurably greater intensity of serving his sentence solo, during the coldest and shortest days of the year.

On the twenty-eighth, however, he wrote, "Although, previous to this blizzard, there had been several days of bright fine weather, but very cold, there has been no sign of the aeroplane. I very much doubt if it will come now. Probably it cannot take off, so I suppose I shall have to wait a few more months until someone can sledge here."

That entry poses several questions. Had Watkins (or Chapman) promised a date by which the Gypsy Moths might try to fly to the station? If so, that vow has escaped the record. In point of fact, another eleven days would pass before the first failed attempt to fly to the ice cap outpost; seventeen days after that before the second likewise unsuccessful attempt. Courtauld could not divine the cloud bank or the headwind that had thwarted the relief flights. "Probably it cannot take off" was the kind of blind guess that stranded refugees concoct to fill the void of unrealized hopes.

At the beginning of his stay, Courtauld had set himself the date of March 15 by which he ought to expect a relief party to arrive. But he had not bargained for the loss of paraffin, or for the fiendish battles he would wage to keep the tunnel exits open. Despite his plucky resolve to put the best face on things, the end of January marked a turning point. He could finally half admit to himself that he was sick of his solitude, sick of the endless drudgery of six observation trips each day, sick of the cold and the monotony. Thus on February 1 he wrote, "I wonder when, if ever, I shall get away from here. Not that I am bored, but I notice that my legs are getting very thin, partly from want of exercise and partly from lack of fresh food I suppose."

Gamely, he kept up his voracious reading, reviewing each tome he finished like a schoolboy tasked with a book report. *Vanity Fair* stood up well on his second time through: "There is something very satisfying in reading perfect English whatever the story." But Thomas De Quincey's *Confessions of an English Opium-Eater* was a harder go: "long-winded and pedantic, but again it is well worth reading for its style."

That same day Courtauld faced a new dilemma. All his paraffin was stored in two big cans. Because there wasn't room for both containers inside the tent, he had stored one of them outside, not far away. But now, with the supply in the first can almost used up, he dug through the drifts outside, slammed by a ferocious gale out of the northwest, but failed to find the second can. The test holes he dug as he searched filled with blowing snow almost as soon as he could start them. At last he gave up, mortified by his failure at what should have been a routine task. "I wish there could be some decent weather," he complained to his diary, while ruefully adding that on the rare clear days the temperature stood at an unrelenting fifty below zero Fahrenheit.

The next day he found the buried can, six feet below the surface. He attributed the triumph to a pair of mittens Mollie had given him. "Finished Pepys," he recorded that evening. "Wish it was continued further."

Through February, Courtauld's diary entries grew less frequent. He kept up his reading, and his dutiful "book reports." Plowing his way to the end of De Quincey, he produced his final verdict: "Do not care for it." But like a schoolboy challenged by his teacher, he added a couple of lines of critique: "Too pedantic although the English is perfect. Also it is plagued with footnotes." He picked up *Tess of the d'Urbervilles*, even though he had read it before. "V. G. especially the first part."

By February 14 he was facing the same problem with the ration boxes that had tormented him twelve days earlier with the paraffin cans. The only box inside the tent was almost empty, but the others outside were buried under what he estimated as ten feet of drift and new snow. That day it was too cold to dig (fifty-eight below zero), and he had an even more pressing problem to solve. He had absentmindedly left the "door" of the new shaft he had dug upward from the side tunnel uncovered, and by now so much snow had piled inside that he had to spend all day excavating it. At the end of that desperate effort, the tunnel was still so constricted "that I can only just wriggle through it."

The tone of the diary entries now becomes flat, with dejection the primary chord—still far short, however, of despair. "Have now been ten weeks alone," he wrote on February 14. "Weather this week has been damnable." Three days later, after a herculean effort fighting the drifts just to escape the "house," he dug for two hours through "concrete snow" until he hit the lid of a ration box. "I got at it and burst it open with the screw-driver. As a last blow found it was one of the ones that had been pilfered of chocolate [presumably by some crewman on the *Quest* on the way to Greenland]. However, thank God, I have got something decent to eat out of it."

That evening he mused again on the prospect of relief by a sledging party. It was not in Courtauld's nature to cast aspersions on his teammates, not even in the privacy of his diary. Instead he wrote, "Hope the chaps have not set out from the Base yet. This weather is impossible for sledging. As far as I can see six months at least of the expedition will be a complete waste of time. All the proposed sledge journeys will go to pot." Then, on February 19, a "terrifying thing" happened. At 5:30 p.m., as Courtauld lay in his sleeping bag reading, "I heard a rushing sound seeming to come from behind and increasing in a second to a roar like an avalanche. It ended with a sudden crash like thunder. I thought at first I was going to be overwhelmed in some sort of whirlwind." But then—silence. At 7:00 p.m. Courtauld went outside to check for damage, but found nothing amiss.

Nonetheless, he was badly shaken. "What it was I cannot think unless some chasm forming in the ice underneath. It sounded most like thousands of tons of snow falling in an avalanche."

Seven days passed before he wrote in his diary again.

* * *

After the second Gypsy Moth was badly damaged on February 26, it took three days to organize a sledging party to head up onto the ice cap to relieve Courtauld. Gino was later criticized for not taking charge of that party himself, and for not assigning the best of the other navigators to that vital effort. But hindsight is foresight, and the essence of Watkins's style was to "teach" leadership to his teammates by putting them in charge of the most difficult journeys. Besides, there was no one he trusted more than Jamie Scott.

As recounted in the prologue, Scott and Quintin Riley got off to a terrible start on March 1, barely cresting Buggery Bank, then, just beyond Big Flag Depot, having one of the two sledges break in half, dictating a return to base. Three days later, reinforced by Martin Lindsay, they had an even worse trial, forced to camp for four days in storm and whiteout well short of Big Flag, with no stove, sustained only by Riley's bag of biscuits and snow melted in cups over a candle flame.

The trio made a third start only on March 9. They got beyond Big Flag in reasonable time, but after that more storms wreaked havoc with their progress. A small anthology of Lindsay's dour comments during these days captures the men's frustration and disgust:

> We only managed a few miles that day as the going was so bad. We said it was the worst we had known, but we did not want to exhaust our superlatives too early in the journey.
>
> During the next six days we only managed to travel twice. We were surprised to get strong gales inland, as in theory the wind is only powerful near the coast.
>
> On the sixth [day] we went a hundred yards, and then the rising gale reduced the visibility to a whip-lash and we decided to camp before we lost each other. . . . When we had fed the dogs, settled down in the tent and dried our clothing, we had done eight hours' work—to cover a distance of a hundred yards.
>
> [In the tent] it was too cold to read with any enjoyment; hands were better kept inside the sleeping-bag. We just dozed and talked—talked of everything from the League of Nations to the price of supper at the Café de Paris.

After those setbacks, the trio suddenly got four days of good weather, during which they averaged just over ten miles a day. By March 15, Scott, Riley, and Lindsay had covered about sixty miles along the flagged route, or almost half the distance to the Ice Cap Station. Those sixty miles, however, had taken sixteen days of brutal effort and two false starts. And the men were quite aware that the party that had established the station the previous August had needed only seventeen days to cover the full 130 miles, despite their terrible struggle pioneering the route through Buggery Bank and the time it took to set up marker flags every half mile. March 15, moreover, was the day that Courtauld had fixed months before as the deadline by which he expected the relief party to arrive.

Courtauld's last February diary entry strikes a new note, as desperation creeps into his hitherto even-tempered jottings. The skies had cleared, but the temperature was down to sixty below, and he was "fed up with going out to the observations every three hours and getting all outer clothes covered with snow by having to wriggle through the tunnel." Then:

> Am down to the last four gallons of paraffin now. If the others don't turn up in three or four weeks, I shall be reduced to cold and darkness. If ever I get back to Base nothing will induce me to go on the Ice Cap again. When the others will come God knows. These gales have made sledging impossible and have raised drifts like small hills even here. What the going is like further out Heaven knows. Judging by when we arrived, when the going here was good, I should think it's absolutely impossible.

In early March, driven by that desperation, Courtauld put down in his diary an entry the likes of which he had never before ventured—an agonized metaphysical tirade against the absurdity of exploration itself.

> Why is it that men come to these places? So many reasons have been ascribed for it. In the old days it was thought to be lust for treasure, but treasure is gone and still men wander. Then it was craving for adventure. There is precious little adventure in sledging or in sitting on an ice cap. Is it curiosity, a yearning to look beyond the veil . . . ? Why leave all whom we love, all good friends, all creature comforts, all mindly joys, to collect a little academic knowledge about this queer old earth of ours? What do we gain?

The worst was still ahead.

TEN

Courtauld in Purgatory

AFTER THE FOUR days of good traveling on the relief mission, Scott, Riley, and Lindsay were stopped cold on March 15 by new storms. They managed to travel at all on only two of the next six days. Lying in their two tents (Scott solo with his equipment, Riley and Lindsay doing all the cooking in the other shelter), they listened to snow and wind pelt the canvas, as they started to worry about the rations meant not only to get them to the station, but to reinforce Courtauld's dwindling store, and yet have enough left over to eat on the way back to base. They agreed to make each ration box last at least a day longer than planned. Lindsay drolly captures the tent-bound mood:

Nothing is more dispiriting than being confined to the tent, eating food that should be taking you farther on your way. You soon finish your little stock of reading material, and are reduced to studying the manual of surveying and navigation. From there it is but a short step to the wrappers on the food tins, and it is amazing how soon you know them by heart. Years and years afterwards you will be able to tell a student of nutrition and dietetics the precise percentage of Pure Protein and Phosphorus in a Highly Assimalable Form contained in Plasmon Powder, and exactly how to make it into a Delicious Snow-White Cream at Trifling cost.

Without the time-set devices to gauge longitude, it was vital for the men to take accurate latitude readings, in order to reach and stay straight along the sixty-seventh parallel; but in temperatures of −28° Fahrenheit this proved fiendishly hard. Whenever Scott set up the theodolite at the noon meridian, he needed to level the horizon with a mercury bowl attached to the device. But the bowl frosted up, and Scott saw only a shimmer of mercury inside. As Lindsay wrote, "To take an accurate observation on a shivering sun would puzzle even the Astronomer Royal." At the same time, Lindsay was struggling with the tripod legs of the theodolite, which sank unevenly in the snow.

When the weather finally cleared, the men emerged from their tents into a temperature of −46° with a headwind blowing in their faces. Even so, they were glad to be moving again. Nearly all the marker flags had blown away, or collapsed to be covered with new snow. The men found not a trace of the big depot at Flag 56, where the invaluable wireless set had been cached the previous autumn. But the sledgers were very gratified to learn eventually that no matter what problems they had encountered with the theodolite, they had deviated from the sixty-seventh parallel by less than one mile. Thus by zigzagging as they approached the Ice Cap Station, they should surely come upon it.

The men were able to travel on five of the next six days, but the nights in the tents had developed their own new trials and discomforts, as a thin sheet of ice—the vapor from cooking and breathing condensed and frozen—coated all four walls. There was no way to move around inside without unleashing a shower of ice flakes. Lindsay greeted the intensifying hardships with his usual wry insouciance:

The cold just has to be endured, actively and consciously, like pain. But it would be absurd to make a great fuss about this, for it is just discomfort in the superlative degree—an ordinary, dull, physical pain which any man who has acquired the right philosophy can cheerfully endure. It cannot compare with any of the agonies of spirit which most people experience more than once in their lives. Nor is it by any means as unpleasant as even the well-known torture of sea-sickness.

As the leader in charge of the relief mission—which more and more began to feel like a rescue mission—Scott shared little of Lindsay's equanimity. Instead, he was assailed by bad dreams. In one, Courtauld arrived back at the base physically unscathed but having been driven mad. In another, a recurring nightmare, Scott visited friends in London but always stopped at the door, too terrified to remove his glove and reach for it, lest he freeze his hand to the handle. This fear was enough to jolt him into consciousness.

The question of what would happen to the Ice Cap Station after they relieved Courtauld wove in and out of the men's thoughts through the long struggle inland. But the burden lay heaviest on Scott. Gino's parting instructions to his friend hung like an albatross from his neck. About the station, Gino had said, "I don't think you'll have to abandon it, though of course you'll have to use your own judgment about that. We could really afford to have only one man there. I don't want you to stay around yourself if the journey home looks difficult." Then he added, "But if it looks easy to get home and the prospect of staying looks bloody—well, I'd rather you stayed yourself and sent the others back."

It was not until March 27—almost four vexed, maddening weeks since Riley and Scott had first set out from base with high hopes of a quick jaunt to the Ice Cap Station—that the three men camped, by their reckoning, only nine miles east of the station. Then, during the next six days, the storms were so fierce that they never gained an inch on the plateau. For six days and nights they lay in their tents, cursing their helplessness, but fearful of plunging into the blizzard and losing their hard-won certitude of exactly where they were.

Scraps of diary entries during those glum, inactive days. Scott: "I feel that we may be quite close so can risk nothing." Lindsay: "It was too cold to read with enjoyment; hands were better kept inside the sleeping-bag." Scott: "Weather equally bloody." Lindsay: "You go to sleep in a tent that is full of warm air; but this soon turns to ice on the sides of the tent. Then you wake up and say to yourself: 'My word, this is hellish—and I'm cold.'" Scott: "Something wrong with the theo[dolite]." Riley: "Things begin to look rather serious."

The men got going again only on April 3, with the temperature at –37°. They had six days' dog food and nine or ten days' man food left provisions that had originally been planned not only to cover the return journey but to augment the nearly empty store at the station. After taking one more latitude reading on the sun, to ensure they were smack dab on the sixty-seventh parallel, they marched west, measuring their mileage with the sledge-wheel. Eight and a half miles on, they camped. Another gale kept them tent-bound through the fourth, but they started again early on April 5.

The men trusted the sledge-wheel readings, which, as earlier tests had showed, varied from true distance by no more than 2 percent. In iffy weather (clouds moving in and out against a milky horizon) they trudged another four miles. By now they ought to have *passed* the Ice Cap Station, but no trace of it broke the level monotony of the plateau. Perplexed, they took another theodolite reading, which revealed a deviation of only three minutes and six seconds from true sixty-seven degrees: "plumb right, though of course we can't count on it," Scott wrote in his diary.

Now the men's vague unease, which had been growing for days, bloomed into full-bore anxiety. The fact that all the recent theodolite readings had confirmed how close the men were to the crucial latitude only deepened the alarm. Where was the Ice Cap Station?

With provisions running disastrously low, Scott decided to kill the first dog, a trouper named August. That evening he shot the husky; then he and Lindsay cut it up and fed it to the other dogs, but only six of them were willing to gorge on their former sledge-mate.

Scott's diary veers between stubborn hope and a kind of nascent panic. Still on April 5 (Easter), he wrote, "If only we get passable weather we ought to find the place." But the barometer was reading ominously low, portending more storms to come. "I doubt if we can stay up here more than three or four days longer," Scott added, "for more than that would probably mean wrecking the teams, which would *muck* up another journey if that's necessary."

The devout Riley was disturbed for several reasons. On Easter morning, he wrote, "I read through the Communion service after breakfast. I wish I was able to make Communion. I do miss it. . . . We can look for the Station for three more days only and then we shall have to return to Base. I pray God we find August first for it will go hard for him if we don't."

What the men could not fathom was how Chapman, returning from the December journey that had installed Courtauld in the station, had reported that the snow walls surrounding the domed tent and the igloos protecting the gauges made up a complex so prominent that it could easily be seen from a mile away. Starting on April 7, the searchers now deliberately struck a course one mile north of the sixty-seventh parallel, traveling almost thirteen miles from what should have been well east of the station to well west of it, but saw nothing. The next day they performed a parallel sweep one mile south of the cardinal latitude, covering thirteen or fourteen miles, and still saw nothing.

Their failure turned their musings toward dark thoughts about ways that Courtauld might have unwittingly thwarted his own rescue. Lindsay speculated, "It was difficult to understand how Courtauld could have come to let the station get so drifted over—unless he had met with some misadventure. A man wintering alone on an ice-cap is hardly a justifiable insurance risk." Riley guessed wildly, "He may walk out, but he will have to cut up the tent. . . . Of course he may have done this [already] and that is why we have not spotted the Station."

As the barometer foretold, storms swept in, and the men were tent-bound on April 8 and 9. Already on the seventh, Riley had faced the worst possibility squarely enough to write in his diary, "If we don't run over the Station I fear something has happened to August and he is dead. . . . It would appear therefore that he has been dead some time for it to drift over as completely as it must have." By April 10, the men had spent twelve days in the vicinity of the Ice Cap Station, on nine and a half of which the weather had kept them tent-bound.

The first line of Scott's diary on April 10 signals the team's surrender: "We have been beaten." The men got off by 8:45 a.m. For the first twelve miles of the voyage home, Scott halted his sledge every three hundred yards or so, stood on top of it, and stared all around, still hoping to spot the station. But the desperation of the return march was now his chief concern. "We will have to hurry for home now," he wrote in his diary. "We have only four days' dog food at full rations for the dogs. At two-third rations it will last six days and we have not too much man food." Lindsay added: "Now it was a race for the Base, and we had the bare minimum with which to get back."

Yet even in their despair, the three men held out hope for one more search effort. The faster they returned, Scott reasoned, the greater were the chances for another relief mission.

The same day, as the men sledged east, Riley's thoughts turned once more to Courtauld. "Poor August's people and Molly," he wrote in his diary. "Unless he has walked out they will have six weeks of suspense at least."

*   *   *

By early March, as Scott, Riley, and Lindsay had finally got past Buggery Bank on the start of their relief mission, Courtauld could no longer ignore his fear that something might be amiss. On March 7, trying to fend off his uneasiness, he wrote in his diary, "I reckon unless something has gone wrong, the relief should arrive between the 15th and the end of the month. One can only trust in God." Already he seemed willing to extend the arbitrary deadline of March 15 that he had set himself back in December, even though by the seventh he had only two gallons of paraffin left and had reduced his food intake to less than a pound a day.

He distracted himself by rereading all three books of *The Forsyte Saga*. ("Extremely good.") But in down moments he dwelt on grim scenarios like the ones Riley and Lindsay had entertained near the end of their own much shorter, warmer stay at the Ice Cap Station the previous September. An epidemic had killed all the dogs; or a crevasse had swallowed the whole team; or the base camp hut had caught fire. . . .

On March 17 another storm filled the shaft Courtauld had dug in the roof of the side tunnel. He tried to excavate his way through the piles of packed snow, but it was beyond his strength. So he turned to the other side tunnel, itself filled with snow. After an ordeal of scraping and digging, he reached the igloo at its end that housed one of the gauges, but it too lay completely buried. He dug straight upward, trying to reach the surface with another vertical shaft, using only his knife to carve away the snow and a biscuit tin to toss it aside. He had almost given up when he suddenly burst into daylight, after clearing five feet of drift. Outside he reveled in his freedom as he sauntered about the snowstricken camp, in sunlight so bright he had to wear dark glasses for the first time. Then he lowered himself back down the new shaft. In an effort to keep the new exit clear, he rigged an empty ration box as a temporary hatch.

On March 22, the inevitable happened. Another fierce storm swept the ice cap, dumping many inches of new snow that drifted heavily over the shaft in the side tunnel. When the storm subsided, Courtauld found that his ration-box hatch was sealed tight under an immovable load of frozen snow. After a futile effort with knife and biscuit tin, he crawled back along the side tunnel, into the domed tent, and slid back into his sleeping bag. In his diary, he wrote, "So I am completely buried. Paraffin has v. nearly run out and things are generally pretty dismal."

At that moment, Scott, Lindsay, and Riley, twenty-two days out of base, had finally gotten moving again after six days of foul weather. They were still at least forty miles short of the Ice Cap Station.

Many men, suddenly entombed in a shelter designed to protect them from harm, would have succumbed to blithering panic. Not Courtauld, with his deliberate rationality, his faith in God, his penchant for seeing the upside of daily tribulations. At once, he wrote down the three sources of immediate threat that could turn his entrapment into a quick death sentence:

1. That the air would become vitiated by reason of there being no possible entry for fresh air, and only an exit in the shape of a 2-inch [in diameter] ventilator in the roof of the tent-house.

2. That the accumulating drift of snow on this roof would crush the tent, now that I could no longer clear it away.

3. That since I could no longer keep a look out, the relief party might miss the Station.

Almost at once, Courtauld satisfied himself as to the unlikelihood of all three. The ventilator tube kept working as it always had: with his nose close to its lower end, he could feel the outside air flowing into the tent. The fear of drifts collapsing the whole domed-tent complex Courtauld dispelled by noting that the drifts had already reached the top of the roof, but showed no signs of piling higher. The domed tent had been sturdy enough to withstand all that strain and weight, and there was no hint of its even starting to buckle. As for the third fear: Courtauld convinced himself that the Union Jack flying on its tall pole was a beacon no search team could miss. "It was clearly futile to get anxious," he told himself in writing, "when by no possible endeavour on my part could I make any difference to the course of events."

In the wake of such catastrophic fears, Courtauld fell prey to his sorrow that he could no longer keep up his daily round of observations—almost a sense of guilt, as if he had let down the team. "A man dislikes changing his habits," he wrote without apparent irony, "and this business of the weather had become a very absorbing habit. . . . Now that I was prevented from doing my job I naturally felt that I was wasting time, and throwing away an opportunity not likely to occur again."

Idle inside his sleeping bag, he turned his thoughts to Mollie. He reread her letters again and again: "It is the only thing left to do that gives me real pleasure." On several days after the March 22 entombment, he put on all his clothes and crawled through the tunnels to scrape away at the blockage with his knife. He still harbored a vague plan that if the relief party never came, he would have to cut himself free and somehow walk (without a sledge) all 130 miles back to the base. But an hour's scraping with the knife, all he could manage each try, made it clear that he would never be able to carve his way out of the snow trap.

One day he summoned the energy to write an extended catalogue of fantasy wishes in his diary. The entry has an almost valedictory tone.

The following pleasures I should like to have granted most, if wishing were any good. One: sitting in an armchair before dinner, in front of a roaring fire, listening to Mollie playing and singing. Two: eight o'clock on a fine summer morning at sea, at the helm of a small boat, a fresh breeze blowing, all sail set, with Mollie and a smell of breakfast coming up to say good morning. Three: just having got into bed with clean sheets and ditto pyjamas. Four: bright autumn morning, eating an apple in the garden before breakfast (an enormous one—kippers, poached eggs, kidneys and mushrooms, cold partridge). Five: getting into a hot bath.

All the while, his supplies of paraffin and food were dwindling alarmingly. None of the diary entries between April 3 and 9, during the days that Scott, Lindsay, and Riley were sweeping back and forth across what they were sure had to be the vicinity of the Ice Cap Station, betrays the slightest hint that Courtauld detected their presence—not even the faintest sound outside the domed tent, unattributable to the normal shriek of wind or slough of sliding snow.

On April 5, Easter—the same day that Riley regretted not taking Communion and Scott insisted that with good weather they still ought to be able to find the station—Courtauld wrote down his most anguished entry yet:

> Now been here alone four months. No sign of relief. Only about a cupful of paraffin left and one or two candles. Have to lie in darkness almost all the time. Chocolate finished and tobacco almost. What a change from last Easter.

But in his hopelessness, Mollie—and his faith—sustained him.

> If it were not for having you to think about as I lie in the dark and can't sleep, life would be intolerable. I wonder what you are doing. If I could be sure you were happy I wouldn't mind. But I trust in God absolutely. I am sure He doesn't mean me to die alone here.

*   *   *

On the journey back to base camp from their failed relief mission, Lindsay and Riley were despondent, exhausted, and scared. But Jamie Scott was in the throes of an existential agony. On the return dash, the men had to kill two of the huskies to feed the others. Another dog gave birth to a puppy, promptly devoured by the other dogs. The men's own rations were meager in the extreme. Yet it took them only six days to reach the Big Flag depot, fifteen miles from base camp. The whole mission, though, had lasted forty days—with nothing to show for it. Only three years later, in *The Conquest of the North Pole*, Arctic historian Gordon Hayes would cite the relief mission of Scott, Lindsay, and Riley as a journey on which "the conditions approached the limits of human endurance."

All the way back, Scott tortured himself with second-guessing. As he later wrote,

> I tried to recall the arguments which had led to my decision. I could not remember them clearly, but I became doubtful of their sound-ness; then certain that I had done the wrong thing. We should have searched until the last dog was dead, and we ourselves were in dan-ger; a life for a life, and I had shirked the issue, had come back with nothing but excuses to throw the responsibility on Gino's shoulders.

Beyond the Big Flag depot, Scott let his partners take the lead, as he was "utterly done." "Alone," he added, "I should have lain down to wait for the strength which might come with daylight." In the darkness of late evening on April 17, the men "slid" down Buggery Bank. "Our loads were pathetically light, but the wretched dogs could not haul them . . . ; so we let them run loose, dumped the sledges, and started in single file to walk the five miles to the Base."

At the tide crack only a mile from the hut, Scott sat down on a stray barrel and asked Riley for a cigarette: "Somehow I had to collect myself before bursting into the sleeping-hut and facing the hail of questions."

But the base camp dogs had heard and smelled their returning brethren, and set off a chorus of howls. Freddy Chapman was the first to come to the door. He stepped outside and shouted, "Who's that?"

The team answered his appeal. Then Chapman cried, "Have you got August?"

Scott yelled back, "No, blast you!"

"There was silence," Lindsay later wrote. "A few seconds seemed to contain all the weariness of a life-time. Then mercifully he said, 'I'll go in and tell Gino you've arrived.' "

Scott's failure to rescue his teammate would haunt him throughout his remaining fifty-five years on earth. He would never go on another expedition. According to his son, Jeremy, "What had happened on the ice cap had shattered his image of himself. . . . [H]e created for himself a personal demon who would cling to his back and reproach him for the rest of his life."

Gino took the bad news in stride. As Scott recounted,

> Watkins came out in pyjamas as we reached the door. . . . He did not seem either pleased or angry, excited or disappointed by what I had told him; merely interested.
>
> He thought for a moment. Then he said: "August must still have food, so he'll be all right if we get there quickly."

After the expedition, Gino was criticized by several outsiders for seeming to be so casual about Courtauld's ordeal. Some said that he himself, as probably the best navigator in the team, should have led the March relief mission rather than entrusting it to Scott. Indeed, that spring Gino was preoccupied with other sub-expeditions besides the all-important central program of the Ice Cap Station. On March 10, only a day after Scott, Lindsay, and Riley had made their third start on the inland relief trek, Gino sent a team of three off on a separate journey. Led by Chapman, the team was rounded out by Lawrence Wager and "Steve" Stephenson. The goal was to use dog sledges to take a route along the edge of the ice cap all the way back to Kangerlussuaq Fjord, at least 300 miles to the north, where Gino's Northern Journey team had left a big cache the previous August-September. And also, along the way, to make the first ascent of Mont Forel, despite Gino's aerial discovery of a whole sub-range of higher mountains beyond Kangerlussuaq.

That journey ended up becoming the least successful in the whole roster of ambitious BAARE discovery jaunts. The sledgers not only failed to get even halfway to the big fjord, but they had to turn back well short of Mont Forel. Throughout the three weeks they spent on the journey, the same terrible weather that was hindering Scott's relief mission thwarted Chapman's threesome, and the sledging conditions were atrocious. The men got a bad scare when Wager fell into a crevasse from which he engineered a narrow escape. The team turned around on April 4, and got back to base camp on the fourteenth, three days before Scott, Riley, and Lindsay limped home.

Now Gino chose Chapman (despite his fatigue after the hopeless jaunt north) and John Rymill as his partners for the do-or-die final jaunt to rescue Courtauld. Rain and poor visibility delayed their departure for three days, but they got off on the afternoon of April 21. If Watkins had seemed unfazed by the failure of Scott's team to find the Ice Cap Station, he was now all business. Scott, who knew him best, thought he understood the psychic stakes that Gino was juggling. "If he should fail to find Courtauld, or if he should find him dead," Scott later wrote, "I believed it would break Gino completely as an explorer. But he had no room for such thoughts." Gino set off on April 21 exuding confidence. Yet, mused Scott, "The game was was still far from won; but at least he would not come back with excuses. Where a life was concerned he would stake everything."

Before he left base camp, however, Watkins had Percy Lemon send a wireless message to the greater world, starting with London and Copenhagen. That announcement, no matter how well-intentioned, would soon become a major thorn in the team's side, threatening the very autonomy on which all the members of the BAARE prided themselves.

The message Gino dictated drily recounted the failure of Scott's team to find the station, and announced the imminent departure of his own team. Courtauld still had plenty of food, Gino asserted. But he added a caveat: "There is always the possibility that Courtauld is not alive, or unwell, in which case Station is probably completely covered [with snow]." No doubt Gino felt compelled to send the message in part because Courtauld's uncle had been the generous benefactor who had saved the BAARE from expiring on the drawing board. But he also anticipated the well-meaning auditors in Europe who might try to leap into action to aid the rescue. Before parting, Gino told Lemon, manning the radio, "to stop any search parties at all costs and to say that everything possible was being done by us."

Gino was naïve. On April 22, the day after he, Rymill, and Chapman started inland, Percy Lemon got a message from the expedition committee in London, announcing that its members felt they needed to send the alarm on to *The Times*, which owned press rights to the BAARE. The next day's headline read, "Anxiety for the Safety of Mr. Courtauld."

The news, of course, spread like wildfire, and with each rephrasing the errors and fantasies proliferated. "Marooned on an Ice Cap," another paper trumpeted, while the *Daily Express* announced "Millionaire's Son Alone on Ice." Yet another paper led with "Danger from Wolves." In rueful retrospect, the BAARE members later agreed that their favorite story was one that appeared in a French newspaper, which reported that "Mlle Augustine Courtauld, the only female member of the Expedition, had spent the winter alone 140 miles from the rest of the party and that every attempt by the men to reach her had been unsuccessful." Yet another paper broke the scoop that Courtauld had managed to send a wireless message to base camp with the curt complaint, "Absolutely without provisions."

As Gino had feared, the expedition committee felt it had to send a rescuer to Greenland to aid in the search. The committee hired an experienced Swedish pilot, one Captain Ahrenberg, at the helm of a larger, faster airplane than the Gypsy Moths, to fly from Malmo via Iceland to

Angmagssalik, then to airdrop supplies to Watkins's team en route and push on to search for the Ice Cap Station from the air.

During the first week of May, six other self-appointed rescue missions sprang into action. Not one of them got so far as the eastern edge of the ice cap. Nonetheless they managed to spread yet another dire rumor: that not only was Courtauld missing, but that Watkins, Rymill, and Chapman had disappeared. "They are believed to be roaming somewhere on the vast ice sheet in the interior of the country. Rescue operations are being pressed forward with feverish activity."

In reality, Gino's team made a good start, reaching the Big Flag depot with ease, as supplies for the rest of the journey had already been cached there. They now had rations for five weeks, stretchable, they thought, to ten or even twelve weeks if necessary. Unlike Scott's ill-fated team, Gino's was favored with perfect weather: clear skies and very little wind. In these conditions, they had no trouble averaging ten miles a day. The men steered by compass and measured their distance by sledge-wheel. Every few days at noon they took a meridian observation for latitude, but, crucially, they also carried a time-set device to gauge their longitude. By May 3, only thirteen days out of base, they were convinced they were within a few miles of the Ice Cap Station.

*   *   *

On the day that Courtauld found himself locked for good inside the domed tent, Scott's team was still many miles short of reaching the vicinity of the Ice Cap Station. Three weeks of utter isolation later, on April 13, Courtauld suffered another setback that seriously sabotaged his will to persevere: "Finished my last pipe of baccy today. There is now precious little left to live for." In his diary, he continued, "Still impossible to get outside. Feet keep on freezing up and have to be always taking off socks and warming them with my hands. Hardly any paraffin left, or candles. I suppose I shall soon be reduced to chewing snow. At present am reduced to a pint of water a day and under a pound of food."

Since he had been entombed, the shelter had become increasingly squalid. The tent walls had finally started to bow inward, always frosted up unless the stove was running, when they melted in a steady drip. All his belongings and his food lay in disorder around his sleeping bag. Since March 22, he had had no recourse but to urinate and defecate in a corner of the tent. The stench became nauseating. On April 20, still a day before Gino's team left base camp, Courtauld lit his last candle and wrote, "Hardly any paraffin. Lie in the dark all day designing the ideal cruise and the ideal meal. Left foot swelling up. Hope it isn't scurvy." Five days later: "Been here alone twenty weeks. Everything running out. . . . What I shall do for drinking water I don't know. Only two more biscuits. . . . Smoking tea as I have no fuel to cook it with." On May 1, with only a little pemmican and margarine and a few drops of lemon juice left, he bravely promised himself to find some way of breaking out of his snow trap and walking with neither food nor sledge 130 miles back to base camp.

Camped, they were convinced, within two or three miles of the fugitive station, Watkins, Chapman, and Rymill had to stay in their tents most of May 4, as another gale sent new snow swirling across the plateau. As soon as it cleared in late evening, the men went off on skis on separate paths to search for the station. As Chapman later wrote, "There were many snow-drifts of extraordinary size whose black shadows were visible for almost a mile, and often we went racing towards one of these, thinking it was the Station." Gino and Chapman returned to their camp at 10:00 p.m., Rymill not until midnight, having, as he explained, "quartered the ground for about 20 miles." None of the three had found the slightest trace of human presence.

If doubt and dread disturbed their sleep that night, none of the three ever admitted as much in print—though years later, Rymill insisted that despite his thorough but futile search on the fourth, he was convinced Courtauld was still alive.

May 5 dawned clear and calm. Before setting out, the men took new observations of latitude and longitude, which yielded the precise result that they *had* to be about a mile northwest of the station. On skis again, each man took a single dog on a leash, "hoping that he would show some excitement when he scented any sign of human beings." They separated to cover more ground, but stayed in sight of one another, no more than half a mile apart.

Chapman:

On reaching the summit of a long undulation we made out a black speck in the distance. It was a flag. Could it be the Union Jack of the Ice Cap Station? We went racing down towards it at full speed and as we approached we saw a large drift on either side of the flag. It was indeed the Station. But as we got near we began to have certain misgivings, The whole place had a most extraordinary air of desolation. The large Union Jack we had last seen in December was now a mere fraction of its former size. . . . Was it possible that a man could be alive there?

A photo one of the men shot moments later tells the whole story. Two dogs crouch wearily on the snow; the head of the third peeks over a low drift. Subtract the dogs, and the photo shows only the top of the tattered Union Jack on its thin pole, the handle and a bit of the shaft of a spade, and perhaps the upper parts of a weather gauge or two. It is obvious how easy it would be to miss those ephemeral signs of human purpose in the immensity of the featureless plateau. And where is the rest of the complex colony of gauges, each with its guardian igloo? Where is the big domed tent, with its tunnel exit and two side tunnels? Instead, nothing but the level sweep of snow stretching to the sharp horizon.

Is it possible that a man could still be alive beneath all those tons of solid-packed drift?

But then, as they skied even closer, the searchers spotted the upper few inches of the two-inch-diameter metal airway tube, still protruding from the drift. The roof of the domed tent must lie just below. Gino approached, knelt, and put his mouth to the end of the tube.

Throughout the eventless day of May 4, Courtauld had kept thinking about the morrow. May 5 was some kind of important date or anniversary, but exactly what, he could not figure out. Though the thought did not occur to him, May 5 would mark the 150th straight day of his isolation at the station. He knew, however, that on the morrow he would burn the very last ounce of paraffin to heat his final warm meal.

His diary later recorded the doings of May 5:

> The primus gave its last gasp as I was melting water for the morning meal. I was lying in my bag after this so-called meal of a bit of pemmican and margarine and had just decided that I should have to start to walk back on June 1st if I could get out, when suddenly there was an appalling noise like a bus going by, followed by confused yelling. I nearly jumped out of my skin. Was it the house falling in at last?

The appalling noise, of course, was Gino shouting down the tube.

> A second later I realised the truth. It was somebody, some real human voice, calling down the ventilator. It was a wonderful moment. I could not think what to do or say. I yelled back some stuttering remarks that seemed quite futile for the occasion. "Hurray!" they shouted. "Are you all right?" "Yes, thank God you've come. I am perfectly fit."

Soon Courtauld recognized the voices of Watkins and Chapman. Still shouting through the tube, his rescuers gave Courtauld a breathless résumé of all that happened on the BAARE during the five months that he had kept his five-month solitary vigil at the station, including the failed search by Scott's party the month before. But soon Gino had smashed a hole in the tent roof and dropped into Courtauld's lair. They grasped hands, "thanking God that the job was done," all while Gino babbled on about the triumphs and setbacks the team had undergone during Courtauld's absence.

Now Chapman and Watkins hauled Courtauld bodily out of his prison. "I found myself pretty weak," he admitted, but then he was able to ski slowly halfway back to the men's camp, before Rymill met him with the sledge on which he was to ride all 130 miles to base camp. In his diary, Courtauld struggled to capture his feelings. "It was all too good to be true. . . . I was and still am unbelievably happy." Later: "It was the best day of my life." And finally: "It is more wonderful than words can express to be free out of that dark place under the snow and to be really going home."

On the way back, anxious to get the good news out to the world, the three sledgers got the most out of their dogs and made incredibly fast time. Happy to be a passenger, Courtauld sat on a pile of sleeping bags and read *The Count of Monte Cristo* as the dogs charged east. But on the second day, Captain Ahrenberg, the Swedish pilot hired by the expedition committee, flew over and dropped food and letters. The reaction of Gino and his partners was annoyance. As Scott later paraphrased their feelings (overlooking the awkward question of who had enlisted Ahrenberg in the first place), "The spirit which had been expressed as 'Damn you, I'm all right,' did not welcome assistance which had not only been unasked but positively discouraged." Worse, it was Captain Ahrenberg who soon broke the good news to the world, not the teammates who had saved Courtauld's life.

On the night of May 8–9, the men slept for only two hours, then covered twenty-six miles to arrive at Big Flag depot in the early night. Instead of camping they pushed on. When they reached Buggery Bank, they threw away the remaining dog food, climbed atop the sledges, and blasted down the dangerous ice track as though it were an Olympic bobsled run. They got to the base camp hut at 5:00 a.m. on May 10.

Bursting through the door, the three rescuers jocularly demanded breakfast. At once they were surrounded by half-awake, ecstatic teammates. They all grasped Courtauld's hand and wrung it ragged. (One would like to think they hugged, but British reticence stops short of committing such emotional excess to the printed page.)

All the men congratulated Courtauld for his astonishing survival, as they demanded that he tell them the whole story. But before he could get the words out, Courtauld sought out Scott, Riley, and Lindsay—and apologized to them, insisting that "he was sorry to have caused so much trouble."

More than a year later, when he contributed a short chapter about his experience to Chapman's *Northern Lights*, Courtauld added a note to the story that heretofore he had shared with no one on the team. It was about a "curious growing feeling of security that came to me as time passed."

> As each month passed without relief I felt more and more certain of its arrival. By the time I was snowed in I had no doubts on the matter. . . . I will not attempt any explanation of this, but leave it as a fact, . . . that while powerless to help myself, some outer Force was in action on my side, and that I was not fated to leave my bones on the Greenland Ice Cap.

Perhaps. Courtauld's faith in God indeed sustained him in his darkest moments. But the diary, immediate and unrevised, may tell the truer story—as on April 13, when he smoked his last pipe of tobacco: "There is now precious little left to live for."

ELEVEN

Asking for Trouble:
Gino's Finale

WITH THE HAIRBREADTH rescue of Courtauld, the BAARE had reached its dramatic climax and happy denouement. The destruction and abandonment of the Ice Cap Station spelled the end to Gino's plan to man a weather post through a full year.* And weeks before that, the even bolder plan to fly a Gypsy Moth (or even two) all the way, by skips and jumps, from Angmagssalik to Winnipeg had had to be scrapped, thanks to the serious damage incurred by the planes and to their limitations even when airworthy.

The great retrospective irony for the BAARE is that it was not the atmosphere above the ice cap that would prove of vital scientific importance, but the billions of tons of ice beneath the surface. Fifty years later, beginning in the 1980s, scientists would retrieve ice cores from the ice cap that ultimately told the history of the last 110,000 years of the climate of the whole earth.

---

* Throughout the eight months that members of the BAARE occupied the Ice Cap Station, all of them wondered just how deep the ice sheet really was. The first answer came from a German expedition, led by the great scientist Alfred Wegener, the discoverer of continental drift, that was exploring the ice cap 300 miles farther north during the same winter of 1930–31. Using dynamite and a seismograph, Wegener's team demonstrated the astounding depth of 8,200 to 8,830 feet of solid ice beneath its central station at 9,850 feet above sea level.

The *Gertrud Rask*—the only ship of the year to stop at Angmagssalik, so keenly anticipated by the Inuit—was due to arrive as early as July 28. Originally, the plan had been for all fourteen men to board the ship for the cruise back to Copenhagen. Nothing would have been more logical than to use the month and a half between Courtauld's return to base and the arrival of the ship than to tidy up some of the mapping projects along the nearby coast, maybe make one more attempt on Mont Forel, then head home with a solid year's worth of discovery and research under the team's collective belt.

But just as Gino seemed to minimize the close call and extended agony that Courtauld had endured, so his mind was aflame with further exploits of discovery—each of which by itself would have amounted to a healthy menu for most of the Greenland expeditions before the BAARE. Gino Watkins was one of those rare explorers for whom no wilderness adventure ever lasts long enough.

On the same day that Courtauld reached base camp, he sent a wireless message to Mollie: "Got back from ice cap today. Fit as an orchestra. Terribly sorry kept you so long without news." Two days later, in a letter to his fiancée, he insisted, "This press rot is all completely obscene. . . . The incident was very mere." Courtauld learned from a telegram sent by his parents that they were not very happy about having "spent thousands" to send Captain Ahrenberg to his rescue. As usual, he blamed himself. "I have been very rude to everybody about this absurd relief," he wrote Mollie. "I don't suppose [my parents] will speak to me again."

Of course it was Gino's wireless alert to the world that had set in motion not only the newspaper-headline frenzy but the rescue efforts launched from England. The very ambiguity of that message—"There is always the possibility that Courtauld is not alive," paired with the order to Lemon "to stop any search parties at all costs"—set in motion the "press rot" and the would-be saviors from Europe.

Throughout the year in Greenland, Gino had cast an envious eye on the locals' prowess at hunting, and his admiration focused on the craft that Inuit peoples all across the Arctic, from Alaska to Greenland, had invented and perfected. By midwinter, it would not be stretching matters to state that Watkins had fallen in love with the kayak. He and several of his teammates had responded to the offers of a few of their Inuit friends to try out their own kayaks, but the Englishmen found it impossible even to get into one. According to Chapman, "The Eskimo has small, rather undeveloped legs which bend backward in the most phenomenal way. Most of the young men could not only touch their toes with their whole palm, but walk along on all fours without bending the knee. Indeed, they could get into kayaks which were only 5 or 6 inches deep."

Instead of giving up, Gino, and then Lemon, commissioned Inuit craftsmen to custom-build kayaks high-decked enough for unlimber Brits to crawl into and paddle. By early spring, most of the other teammates had followed suit, caught up in the fun of Gino's new "hobby."

Except it wasn't a hobby. As seal-hunting formed the core of Inuit existence, the techniques employed varied hugely with the seasons. In the winter, with ice spread over the coves, seals had to be harpooned after patient vigils over breathing holes or shot in the rare open leads. In spring, the seals crawled up onto the ice to bask in the sun, and a skillful hunter could stalk them and even club them to death. But in summer, with so much open water, only hunting them by kayak worked. And the summer kill could ensure the existence of a whole community for another year.

Gino was determined to master the art of seal-hunting by kayak, not as a *jeu de sport*, but as the essential component of the last of his seven mini-expeditions through the Greenland year—the one that rivaled the Ice Cap Station as dearest to his heart. Most of his teammates might be content to call it quits in late July and board the *Gertrud Rask* bound for home, but Gino had cherished from the start a flamboyant finale to his *annus mirabilis*: a one-way journey by whaleboat and kayak all the way south along the coast to Cape Farewell and on to the nearest settlement on the west coast. Along the way, he would make the closest scrutiny yet (mapping each detail for much of the way) of the shoreline that had bedeviled Graah and Holm and every other European mariner since 1885.

If Gino could perfect the art of seal-hunting by kayak, he—and whatever teammates could be cajoled into joining him on such a truly risky journey—could revel in the achievement of living off the land in the same way that the native master hunters had done for centuries. If not . . . well, then, they might be thrust into one more nasty survival ordeal.

Gino's fascination with the Inuit kayak soon surged through the veins of most of his teammates. It's striking today to observe how the same young Englishmen who remained tone-deaf to every echo of the spiritual life of the East Greenlanders could open their eyes to the genius that produced their most remarkable invention. As Chapman rhapsodized, "The kayak of the Angmagssalik Eskimos is not only a wonder of efficiency, but a veritable artistic triumph. It is the perfect canoe. Each detail has evolved till it has reached perfection. The kayak, like a racehorse, is a thing of infinite beauty."

So much for aesthetic appreciation. But sometime during the late winter, Gino "realized that it would be impossible for us to hunt seals in the summer unless we learnt to use a kayak." Why, one might ask, if all that remained of the BAARE was to tidy up the loose ends and wait for the *Gertrud Rask*, would it be important for the men to learn to hunt seals? Why, unless Gino's obsession with "going native" hadn't somehow spread like a virus to most of the team?

Along with admiring the Inuit kayaks, the BAARE men had marveled through the year at the hunters' skills they served: natives "throwing their harpoons with consummate grace" and "dressed in waterproof coats, rolling the kayak right over in the water." It was evident early on that if an Englishman was going to try to stalk seals from a kayak, he had better learn first how to stay alive in that fragile-looking sealskin boat.

For young men steeped in the competitive cauldrons of public school, Cambridge, and the military, the sheer challenge was appealing. In Chapman's appraisal,

> Europeans had learnt to go in a kayak before; in fact Nansen and Johansen had depended on them in that memorable journey to Franz Josef Land after the *Fram* had been frozen into the North Polar Ocean. But it was generally thought impossible that a European could learn to hunt seals from a kayak, or to roll it in the Eskimo fashion.

Thus began the BAARE men's crash course in "the art of kayaking" (as Chapman titled the most beguiling chapter in *Northern Lights*). At first, it was a comedy of errors. The tutees found it impossible to get into their kayaks in the water, so instead they mounted the craft on land and suffered the ignominy of letting their teachers carry them down to the water. Once launched, the men found the kayaks frighteningly unstable. Chapman: "At first it felt most unsafe, very like trying to ride a bicycle for the first time. You wobbled one way, and then went too far over the other way trying to correct your balance, and finally lost it completely." After a little practice, "we could gingerly paddle along, being very careful round the corners, and stared fixedly straight in front, for the least turn of the head started a wobble."

Eventually eleven of the fourteen men ordered custom-built kayaks. Having read ethnographic accounts from other parts of the vast Arctic, and later having watched west coast kayakers, where the ancient craft were already being supplanted by motorboats, Chapman concluded that "the Angmagssalik Eskimos . . . are probably the most accomplished kayakers in the world."

Before attempting any seal hunts, the eleven BAARE kayakers needed to learn how to deal with the omnipresent threat of a capsize.

For months the men had watched Inuit men practicing the lifesaving technique called the "Eskimo roll," first in astonishment, then in envy. The most skillful of the natives could right their kayaks not only with a paddle, but with a short throwing stick or even an arm alone. For Gino, who had shunned all sports at Lancing and Cambridge, this was a challenge as thrilling as mountain climbing. As he told the RGS audience the next December, "Once a man can do this roll with absolute certainty, it means that he is safe to hunt alone, as he will be able to get up again if he is knocked over. I knew that I should have to hunt alone on the coast journey, and I saw that it would not be safe unless I could learn to roll the kayak. It turned out not to be as difficult as it looked."

As usual, Gino downplayed his virtuosity. Chapman's first attempts to roll ended with him thrashing in a panic, upside down in the water, before his Inuit lifeguards hauled him upright with their paddles. Eventually, seven of the eleven BAARE kayakers learned how to roll, "but Watkins was the only one who learnt to do it with the throw-stick or the hand alone."

Just as the kayak was the ultimate refinement of Inuit craftsmanship, so the technique of hunting seals from such a boat stood as the culmination of centuries of struggle against an elusive prey. Rifles had been introduced to the east coast natives as early as 1890, but the skill had been developed for hunting with harpoons, and even in 1930, the hunters were still masters of the traditional style—for one crucial reason.

Because a seal, once wounded, immediately dives deep, a simple harpoon embedded in its hide would disappear with it. So the Inuit had learned to craft a long wooden shaft tipped with a tapered point made of walrus ivory. That point was hinged to the shaft in such a way that it would break loose once it hit its mark. The seal would then head for the deep with only a six-inch dart point stuck to its body. But that point was tied to a forty-foot-long cord made out of sealskin thongs. Neatly coiled in the deck of the kayak, this cord was whipped loose by the frenzy of the wounded mammal. The other end of it was attached not to the hunter or to the kayak, but to an inflated bladder, itself made of sealskin with all the orifices patched up. Thus the bladder, as soon as it was yanked loose from the kayak, acted like a bobber to indicate the whereabouts of the submerged seal.

Sooner or later, the seal had to come up for air, at which time the hunter tried to finish the kill with fierce thrusts of a short lance. But the job was still not finished, for a dead seal would quickly sink—unless the hunter instantly plugged the wound or wounds from which the animal's blood gushed out with bone pegs (an assortment of which he carried in his kayak), rendering it still floatable. At last, at the end of this grueling operation, the joyous hunter would tow the massive seal back to the shore.

So deft were the best hunters at this extraordinary procedure that a handful of men had been known to slay as many as forty seals in a single day.

The problem with using a rifle alone was that there was no way the wounded prey could be retrieved. So an Inuit hunter killed with a rifle only when he was close enough to snag the carcass before it sank. If the seal started to sink, the hunter quickly used his harpoon to secure the catch.

Even the art of throwing a harpoon was a highly evolved skill. A man sitting in a kayak and simply throwing the harpoon with the limited force of his arm and upper body could seldom summon the force necessary to pierce a seal's thick skin. So the Inuit used a throwing stick—a short launcher with the harpoon itself nestled in a groove. The tandem weapon multiplied the power of the harpoon severalfold. It's the same kind of weapon used for millennia by native hunters all over the globe, called by archaeologists the *atlatl*, until it was superseded by the bow and arrow. (That same throwing stick served a capsized hunter in extremis if, having lost his paddle, he could use it to flip his kayak upright.)

As he was learning how to hunt from his Inuit masters, Chapman was stunned to witness their skill with the harpoon. "The Eskimos' aim is amazingly accurate," he wrote, "and to keep it so they are continually practicing. If a man goes out in his kayak to catch cod or collect seaweed, he will throw his harpoon every few minutes as he paddles along."

The whole art of the hunt was predicated, of course, on getting close enough to a seal in the brief moment when it surfaced for air to reach it with a harpoon, for the mammals were so skittish that the slightest unguarded movement by a hunter would send it diving for cover. One aid the natives had invented was a small white screen mounted directly in front of the kayaker's head, which, though it severely limited his vision, completed the illusion for the seal that the amorphous whiteish blob floating nearby—kayak and screen—might be just another stray lump of ice.

Out learning with his tutors, Chapman was amazed once more to see how an Inuit in his kayak could freeze stock-still the instant a dark head bobbed into sight, then paddle so fast once the seal had dived "that the back of the kayak is almost forced under water." What's more, the hunter could predict by the kind of seal that appeared roughly where it was likely to surface again minutes later.

It's a pity that Gino left only the cursory account in his talk to the RGS about his own apprenticeship as a kayaking seal hunter, and that he so understated his achievement in getting up to speed in that difficult art in only a matter of weeks. Martin Lindsay, in *Those Greenland Days*, adds a few wry comments. It turns out that almost all the custom kayak-building for the Brits was performed by the esteemed elder, Nicodemudgy, the man who had killed ninety-five polar bears in his long, adventurous life. "He charged Kr. 100 (about £5) for each," Lindsay wrote, "and as the only opportunity he has of dissipating this fortune is buying his simple requirements at the Angmagssalik store, he is now well-established as the richest man in East Greenland."

As one of the two biggest men on the expedition, Lindsay may have found kayaking a particularly intimidating game, even with his own raised deck to accommodate his gangly body. "*Kayak*-ing is not a pastime to be lightly undertaken," he warned. "Should you roll over before you have acquired the knack of rolling up again, you will stay upside down and give joy to the fishes."

Quintin Riley, in fact, suffered a close call in February, before the ice was out in the cove near base camp. Chapman had shot a seal from an ice floe, and Riley eagerly set out in his kayak to retrieve it. A little over-zealous, he dug his paddle too hard and flipped the kayak. That day there were no Inuit lifeguards in attendance, but Chapman just managed to grasp the tip of the kayak from the edge of the floe and haul his teammate to safety.

In any event, by the end of June, Gino had announced his plan for the BAARE's grand finale. Only six of the men would sail for home on the *Gertrud Rask*. The eight others would stay on to perform three different marathon journeys. All three would be one-way trips, ending on the west coast. And though all three were Gino's pet projects, so seductive was their leader in conveying the challenges and achievements still waiting to be claimed that his seven volunteers, no matter how homesick they may have been after eleven months in Greenland, entered into their pacts with their faint misgivings far outweighed by the passion for one more plunge into the unknown.

* * *

In the meantime, between May 6 and June 9, Wager, Stephenson, and Bingham had made a second attempt on Mont Forel. It greatly improved on the first attempt by getting to the mountain after a trek of 176 miles. But 500 feet below the summit, Wager and Stephenson turned back in the face of a steep, icy dome that Wager thought too risky to climb. Forel's first ascent came five years later, at the hands of a Swiss expedition led by the crack alpinist André Roch.

Two of the three components of Gino's grand finale were to be complete traverses of the ice cap, one on a southwest diagonal aimed squarely at the west coast village of Ivigtut, the other (and more logistically complicated) one, a gently rising west-northwest vector that would pass well to the south of the abandoned Ice Cap Station but arrive at another west coast village called Holsteinborg, a little north of the Arctic Circle.

By 1931, such crossings were no longer really the exploratory breakthroughs they had been for Nansen in 1888. Indeed, Gino seems to have advanced only the excuse for carrying out the pair of traverses. They fell under the excuse of "find[ing] out as much as possible about the height and meteorological conditions of the interior of Greenland."

The Ivigtut traverse was really born of Watkins and Scott's disappointment at failing to complete the "Southern Journey" the previous October. By the summer of 1931, with both the Ice Cap Station and the dream of the Arctic Air Route in shambles, any scientific rationale for the journey was window dressing. Instead, the motivation reverted to old-fashioned adventure and "discovery"—not of scientific truths but of land no one before had seen.

Scott wrote the chapter in *Northern Lights* devoted to the Ivigtut traverse. "I like to imagine that the Ivigtut journey was my own, both in plan and development," he began; but then he felt the need to qualify: "Actually, one realizes that, as usual, Gino Watkins was its modest father, while oneself was no more than the casual mother who schemed and suffered in its development." Scott wanted a team of three, with three sledges each hauled by nine dogs. "For the personnel," he wrote, "Martin Lindsay was an obvious victim," since the two men had gotten along well on previous jaunts, and they "knew exactly how to work together." Lindsay was only too glad to volunteer. But as Scott cast about for a third member, "Steve" Stephenson abruptly stepped forward, even though he was fresh off the grueling, month-long attempt on Mont Forel.

The men packed rations for six weeks, which Scott deemed close to the limit for an unsupported ice cap trek. The distance to be covered—

450 miles—was half again longer than what Nansen's team had faced. Even if they could travel every day of the six weeks, that would still mean averaging ten miles a day. To make the loads lighter, Scott packed full rations for the dogs, but only two-thirds rations for the men, reasoning that in a pinch they could kill dogs for emergency food.

Ideally, both traverses would have started out on July 1. But the limited supply of dog pemmican in camp, and the dilapidated state of most of the sledges, meant that one of the crossings would have to be delayed until the *Gertrud Rask* reached Angmagssalik with more dog food and new sledges. In the end, the Holsteinborg team did not get started until August 13.

Anticipating an all-out push against time and weather to reach the west coast, Scott's trio stumbled into a sledging idyll. No storms halted their steady progress. With at least partial daylight round the clock, the men settled into a 9:00 p.m. till 7:30 a.m. sledging routine, when the snow was firmest, punctuated by a leisurely break at 1:00 a.m. while they pitched their tent, brewed up cocoa, and munched on biscuits, marmite, and chocolate.

After all the agonies of the previous treks the men had endured, they took a giddy delight in the ease and comfort of the Ivigtut crossing. "In the tent we ate our meals lying about carelessly instead of wrapped to the neck in a sleeping-bag," wrote Scott. "Very often we slept naked on top of our down bags." Hoping to cover ten miles a day, they easily averaged more than fifteen. The whole journey took only twenty-eight days, rather than six weeks. Writing up his chapter after the expedition was over, Scott feigned narrative disappointment: "The journey was successful to the extent that there were no adventures, and therefore as a story it is dull."

There was only one truly grim episode—a denouement the men had dreaded from the start. In Ivigtut, the locals raised cows and sheep. As Scott explained, "Huskies and domestic animals cannot live together, and we had been told that we must kill our dogs." On the western edge of the ice cap, the men prepared for the gruesome deed. In his own account, Lindsay saluted the dogs by name, for each had assumed an almost human character: "In Akiyak and Songarywadi we had two wise old men—asthma-ridden lawyers' clerks, meagre and poorly paid. . . . Then we had Owgoot, the scallywag, and Archie, the cockney. . . . Kernek would have been a barber's assistant, who played left-half for the local football team."

The huskies had been raised on the west coast, where Scott had purchased them a year and a half before. Now, almost within smelling distance of that coast, they had only minutes left to live. Lindsay estimated that each dog had pulled a sledge for the BAARE about 1,500 miles. At midnight, the men gave each husky a final treat of ten pounds of pemmican, "as a sort of prisoner's breakfast," in Scott's words. Then they tied each dog to a rock, and Scott walked up to them one by one with his rifle and shot them through the brain. "I sat for a while on the moraine and made my peace before crawling into the tent, where there was no privacy," he wrote. "I read [Milton's] Lycidas and went to sleep."

Three days later the men were greeted by the chief engineer "in pyjamas" on the edge of Ivigtut. That man took the bedraggled explorers to the director of the mining community, whose first words were, "Ah! you would like a bath." Lindsay later wrote, "The joy of that bath, the first for twelve months, will live long in our memories." But Scott closed his own account: "Already the Ice Cap had become unreal."

* * *

From the start, the Holsteinborg traverse was to be a two-man job. John Rymill and Freddy Chapman were slated for the journey. They were the most likely pair: Rymill, the expert navigator who had plotted the original route to the Ice Cap Station and determined its exact position, as well as being probably the physically strongest man in the team; and Chapman, who had taken the lead on more excursions than anyone else except Watkins. But in May Chapman had developed a swollen gland in his neck, and when it got worse rather than better, "Doc" Bingham strongly urged that he take the *Gertrud Rask* back home and get the ailment treated in London as soon as possible.

It was Wilfred Hampton, the engineer and pilot from the Air Force Reserve, who took Chapman's place—enthusiastically, as he had spent too much of the year tending to the temperamental Gypsy Moths and trying to repair them after their smash-ups. Even on the drawing board, the Holsteinborg trek loomed as a more complex operation than its Ivigtut cousin, for it was known that a hundred-mile gap of lowland hills, ridges, and streams loomed between the coastal village and the western edge of the ice cap. So Gino concocted a hybrid attack. On their two sledges, the men would carry a pair of kayaks all the way across the polar plateau. Where the ice cap ended, they would abandon the sledges (and the dogs) and use the kayaks and their own feet to navigate the lowlands. The overconfidence of the plan, it would seem, should have been glaringly evident, for the men intended to kayak down any streams that were too deep or swift to wade. This despite the fact that all the kayak practice throughout the spring and early summer had been undertaken on flat seawater.

Rymill and Hampton got off to a bad start beyond Buggery Bank, for in August the crevasses, crisscrossing their path, had melted open. At one point, six dogs fell simultaneously into a crevasse: they were rescued only by a dicey maneuver when a roped Rymill climbed down into the slot, disentangled the harnesses of the terrified canines, then tied them in one by one so that Hampton could haul them to the surface. Shortly thereafter one of the two sledges got wedged sideways across the mouth of a deep crevasse, saved from plummeting into the depths only by a protruding shelf on which it caught and stuck. Only another delicate and risky process saved the day, as the men perched on the edge of the fissure and unloaded the cargo piece by piece before wrenching the empty sledge to safety.

The vast bulk of the supplies for the Holsteinborg crossing had been cached beforehand at the Big Flag depot, fifteen miles out. But now, absurdly, Rymill and Hampton could not find that well-marked outpost. Only after sweeping back and forth and recalculating positions by taking back-bearings on the coastal mountains, still visible on the eastern horizon, did they stumble upon Big Flag, "on a strip of ice so sunken between two large crevasses that it was invisible from more than 30 yards."

After that, the men sailed blithely across the relatively crevasse-free plateau. With ten dogs per sledge, compared to the Ivigtut team's seven, they made wonderful progress, averaging twenty-four miles a day for ten straight days. By September 4, twenty-three days out from base, they had covered 300 miles. Only forty miles of ice cap still separated them from the lowlands. And here, Rymill and Hampton's troubles began in earnest.

Meanwhile Gino had put together the three-man team for what he called the Open-Boat Journey, a 600-mile nautical crawl along what was arguably the most treacherous coast in Greenland, buoyed by his bold ambition to support the effort in large part by living off the land, thanks to his newly mastered skill at hunting seals from a kayak. Percy Lemon, who had spent almost the entire year chained to the base camp radio, was rewarded with an invitation to join that he gladly accepted. But Gino still needed a third man.

If any member of the BAARE had earned his place on the *Gertrud Rask* for the cruise back to Europe, it was August Courtauld. As Scott drily put it, the man "had so obviously exceeded trade union hours on the Ice Cap" that a quick ticket home seemed his inalienable due. Since he had been rescued from his snowbound tomb 130 miles from nowhere, Courtauld's every thought had been directed toward Mollie, to whom he wrote every few days.

But something strange (though entirely characteristic) came over Courtauld's spirit during June and July. As if reborn into the animate world, he wandered around the base camp hut marveling at the seasonal revival of the frozen world. In his diary, he wrote,

> The country is very lovely just now. One walks in valleys deep with moss and grasses, with thick patches of harebell in full bloom and saxifrages of many sorts. All round are the soaring peaks standing clear against the blue unclouded sky. . . . When we feel hungry the Doc [Bingham] goes out with his rod and brings back a fish which is soon sizzling in the frying pan.

In his darkest moments inside the domed tent, Courtauld had vowed never to go exploring again, but instead to study the history of exploration from the comfort of his armchair with a blaze in the nearby fireplace. But now his father was badgering him by telegram to get a real job when he returned home, since the family fortune had suffered a bad downturn in the Depression. In August Courtauld would turn twenty-seven. Was there to be no more adventure in the future if he heeded his father's recommendation of "a life in insurance"?

Gino's plan for the Open-Boat Journey was no secret among the team. When Courtauld first heard of it, he thought the scheme was insane. An eighteen-foot whaleboat with a three-horsepower outboard engine seemed to his pragmatic mind a recipe for failure: fuel alone would fill so much of the boat that there would be hardly any room left for food. As for Gino's vow to fill the larder with hunted seals, Courtauld was skeptical. It was only when Gino, swayed by the logistics, decided to take a second whaleboat that Courtauld relaxed his objections.

Gino sensed his teammate's dawning interest, and so, at the last minute, he invited him to come along. Despite his misgivings, Courtauld could not say no. The powerful magnetism of the leader had worked its spell. No one else, he later admitted, could have undercut his desire to head home on the *Gertrud Rask* and get on with the rest of his life with Mollie. Gino, Courtauld swore, "could not only lead men by the force of his personality but with his charm inspire them to do more than ever they knew was in them."

He was anguished about postponing the reunion with Mollie. "I feel a frightful rotter about it," he wrote her, "but I know what you would have me do in your heart of hearts. You wouldn't have me run out before the show is over." And he offered the explorer's standard disclaimer to worriers back home that there was nothing in the least bit dangerous about the upcoming journey: "Various things may go wrong, but they can only cause delay and not disaster."

Delayed by pack ice, the *Gertrud Rask* reached Angmagssalik only in early August, the base camp fjord on August 12. Scott, Lindsay, and Stephenson had long since set off on the Ivigtut traverse, but for the rest of the team, the ship's arrival heralded a poignant parting of the ways. Chapman, Wager, Riley, D'Aeth, Cozens, and Bingham boarded the *Gertrud Rask*, while Rymill, Hampton, Lemon, Courtauld, and Watkins stayed on to prosecute the last two Greenland journeys.

The next day Rymill and Hampton set off on the Holsteinborg crossing, and two days later Gino, Courtauld, and Lemon headed south in their two whaleboats. Gino and Lemon were saying goodbye to their Inuit paramours, Tina and Arpika, but the bonds between the whole team of young Englishmen and the Inuit community had grown throughout the year. According to Courtauld, as the whaleboats headed out of the cove, "Men and women sat on the rocks beside the water of the fjord and wept while we left them, probably for ever."

Gino had admitted to Scott, as his best friend headed up onto the ice cap in early July, "I'm looking forward to this coast journey more than I have to anything in my life." Yet originally the skepticism of his comrades had been general. As Courtauld later wrote, "Although most of the members of the expedition called it uncomplimentary names, they all, in the end, said, 'Well, Gino has always got away with it, so I suppose he will this time.'"

The division of labor on the Open-Boat Journey was explicit. Lemon was to be in charge of mapping, maintenance for the flimsy outboard motors, and trying to use a portable wireless set to communicate outside. Courtauld's job was to navigate by astronomical observations and to keep the two boats in general repair. Gino's sole responsibility was to hunt for seals to keep the men from starving.

Yet with his usual meticulous attention to detail, Gino had built in to the logistics the gear to carry out various Plans B if the main program failed. Atop the boats were loaded three kayaks and three hand-sledges with man-harnesses (the latter in case the sea journey had to be abandoned in favor of a man-hauling trek across the southern end of the ice cap). The boats also carried guns and rifles, a hefty supply of ammunition, and the tools that would be needed to build a hut if, in the last resort, the men had to winter over.

Among European explorers, only Gustav Holm, during his epic push in 1883–85 along the southeast coast to Angmagssalik, had closely navigated that rugged shoreline before Gino's Open-Boat Journey. Holm had made a map, on which Gino hoped to improve, but only along the 150-mile stretch from base camp to Umivik, an ancient settlement just north of the sixty-fourth parallel. From there on, across the 450 miles of more dangerous coast, the men would travel as fast as they could, in order not to get trapped in the gathering early-winter ice pack. The goal was Julianehåb, the first major village up the west coast.

Gino had persuaded old Nicodemudgy, the veteran who seemed an expert at everything Inuit, from killing polar bears to custom-building kayaks for the unlimber Brits, to winter over with his clan at Umivik, repairing one or more of the ancient stone houses vanished predecessors had built. Three weeks before Gino's departure, the patriarch set off in a twenty-eight-foot umiak loaded with eight relatives, including his third wife. As the deep fjord at Umivik was renowned for its excellent hunting, Nicodemudgy had no qualms about living off the land there. He also carried twenty-seven gallons of petrol to supplement the cargo of fuel Gino's team needed for the whole trip. That supply had been carefully calculated, with the result that most of the boat space was filled with petrol cans, leaving little room for food.

The first nine days of the journey, in perfect weather, went smoothly, although Lemon's experiment in saving petrol by towing the second boat behind the first backfired, when the trailing craft veered in wayward directions or bumped rudely into the lead boat. The whaleboats, purchased from the *Quest*, were pretty beat up, and despite prophylactic improvements like reinforcing the metal sheathing on the prows, they regularly sprang leaks.

On August 24 the men made camp on a barren little island called Pikiutlek, a little more than halfway to Umivik. There they were stranded by fog and sea ice for a week, during which time they renamed the island Pigsty. In glorious weather they finally got off again, and pulled in to Umivik at dusk on September 1, where they were joyously greeted by Nicodemudgy's family.

The Inuit, according to Courtauld, "were surprised we had so little food with us, as they refused to credit Gino with the ability to hunt seals in his kayak." To prove them wrong, he bagged two seals the first day at Umivik and two the second. He had perfected his own style, which paired shotgun (rather than rifle) at thirty yards or less with harpooning at close quarters. On one of the kills, he had a close call, when the suddenly diving seal dragged the line to the bladder across the kayak itself and the boat flipped onto its side, holding Gino's head and chest underwater for uncomfortably long seconds.

None of the BAARE men had any illusions about how dangerous hunting seals by kayak was, even for the Inuit. Lemon insisted the attendant death rate "among the Eskimos between the ages of 18 and 25 was the highest in the world." Gino later told the RGS that "about 90 per cent of the Angmagssalik Eskimo men end their lives by drowning in the kayaks through being overturned by an attacking seal or walrus, or by a wave" (though the 90 percent figure was an exaggeration).

As the perfect weather held, the men spent a week in Umivik. In his account of the journey in Scott's biography of Gino, Scott gilds the layover as a welcome interlude: "We spent some pleasant days at Umivik, hunting and telling tales with our great friends the family of [Nicodemudgy]."

But privately, he was furious. Gino "said we were going to stay here two days and we have been here five already," he wrote in his diary. "It is very pleasant but I want to get home." They were squandering the perfect weather, while Gino, in no apparent hurry, showed off his hunting prowess to his Inuit friends. Lemon, who wrote the chapter about the Open-Boat Journey for *Northern Lights*, voiced no annoyance over the stay, instead recounting his gratification that for the first time he had made radio contact with Angmagssalik.

But the week at Umivik sparked a chain of tensions and quarrels that would plague the rest of the journey—the closest the BAARE team had come in its Greenland year not only to outright conflict but actually to fearing for their survival as a result of Gino's quixotic decisions.

From Umivik, it was 350 miles to the first tiny settlement on the west coast, 100 more to Julianehåb. Lemon optimistically foresaw "seven days' travelling, perhaps, if all went well"—ten days at a safe guess. The outboard engines were running well, and the men could see no pack ice ahead to block their way.

But the weeklong sojourn at Umivik had used up the good weather. On the first day onward, both motors started misfiring, so the men camped early in hopes a rest would solve the engines' idiosyncrasies. That night a rainstorm came in on a strong wind, and by morning the sea was an angry chop. As the balky flotilla pushed on, the rain only got worse. On the third day out of Umivik, the fight to keep going turned into a debacle. For two hours, the men pulled as hard as they could on the starter cords, but neither engine would spark to life. At last Courtauld got the weaker engine (three horsepower, versus the newer engine's four) firing fitfully. The men set out with Courtauld in the smaller boat towing his two teammates in the larger. Suddenly the motor stopped dead. In the surging waves, the men eyed the dangerous sea cliffs to the right, a new batch of "towering" icebergs on the left, as the boats drifted helplessly on the wind and the current. Courtauld took out the spark-plugs and tried to clean them, even as every wave filled his boat with water. Soon he shouted over the wind, "Someone must come and bail!" Gino leapt into the smaller boat, but as he bailed, he saw that a bucket-ful of water sloshed in with each wave, inundating the boat faster than he could empty it. Meanwhile the boats were drifting unchecked toward the nearest iceberg.

The men solved this life-threatening emergency only by the most desperate action. Lemon threw overboard the seal meat the men had loaded for rations to get to the oars the men had carelessly buried beneath. Watkins tossed his kayak overboard, only to see it flip and drift facedown. Miraculously, Courtauld got the engine running on only one cylinder. The men steered away from the menacing iceberg and sought a landing place. At a snail's pace, Courtauld towed the crippled tandem three miles toward a plausible cove on never-inhabited Skoldungen Island, until a monster wave doused the engine for good. The men rowed the last half mile.

The cove happened to have a sandy beach, a great rarity on the southeast coast, which allowed the men to pull the whaleboats com-pletely out of the water and to find a decent campsite in the partial shoulder of a big boulder. But here they were marooned for four days. It rained "in torrents," Lemon reported, sending little streams under the tent that soaked everything; but at least it "saved us going outside for [drinking and cooking] water."

In seven days out of Umivik, the team had advanced fewer than a hundred miles—a small part of the 350 to the first settlement of the west coast that Lemon had blithely guessed the men might cover that same week.

On Skoldungen Island, Gino seemed to fall into real depression, a state his teammates had never before witnessed through the whole year in Greenland. The sodden camp on the beach would have depressed the most cheerful camper: the bilge that had filled both boats had soaked everything, including the men's film, books, and writing materials, as well as most of their clothing and food. An emergency bag of oats lay stuck together in a gooey mess. In addition, the men were out of tobacco, and they were short on matches. "We all asked ourselves why ever did we start on this absurd journey," wrote Courtauld.

The radio, too, was soaked. Lemon could not receive with it, and doubted whether he could broadcast. So Gino decided to abandon the device, with its considerable weight, there on the beach. On the chance that a last message might be picked up somewhere, he had Lemon send out word that the men were considering abandoning the boats to man-haul across the ice cap, or alternatively, to winter over in place. To forestall any attempts to rescue them, Gino purposely gave out no indication of where the men were.

An emergency wintering in the Far North had been a fantasy Gino had nursed since the Edgeøya traverse in 1927: after all, Nansen and Johansen had pulled it off in the winter of 1895–96 in their do-or-die retreat from their attempt to ski to the North Pole, huddling for eight months in a turf hollow they dug with a walrus shoulder blade. But for Lemon and Courtauld, wintering on Skoldungen Island seemed to loom as a death sentence.

All the while the men waited out the fiercest rains they had seen during the whole year, Lemon and Courtauld tirelessly tinkered with the outboard motors. At last they got one to start. So the men decided to leave one boat on the beach and complete the journey with other, which they had named the *Narwhal*. Without the dead weight of the radio, they could just barely fit all their goods into a single boat. In improving weather, they set out again. From the *Narwhal* Gino managed to shoot a number of guillemots and seagulls and a single duck to supplement their depleted stores. During several good days the men put another eighty miles of coast behind them, before having to camp in "a little crack in the cliffs" as the rainstorms returned. More ominously, the sea ahead was thick with bergs and floes, harbingers of the winter that lay just around the corner.

Courtauld summed up the men's predicament:

> We were now at the most critical stage of the journey, for it was too far to go back at such a late season of the year. The mountains were impossible to get over with hand-sledges for crossing the ice-cap, while the hunting on this part of the coast is not good enough for wintering.

Lemon's account of the Open-Boat Journey in *Northern Lights*, while full of vivid details, gives not the slightest hint of the conflicts among the men that were escalating day by day. Nor does Courtauld's narrative in Scott's biography of Watkins. Both men observed the age-old code in exploration writing of leaving the dirty laundry out of sight of the reader. But Courtauld's diary sheds some light on the acrimony.

Evidently, Lemon and Watkins didn't share Courtauld's judgment that the hunting near their cliff-crack camp was too poor to sustain three men through the winter. But "Gino seems to want any excuse for staying, and Lemon backs him up. I have said nothing yet, but my opinion is never consulted." Courtauld even suspected that the desperate travel in storms with leaky boats had caused Gino—never a happy sailor—to lose his nerve for coastal navigation. But what stung Courtauld with anguish was the prospect of delaying his reunion with Mollie for another whole year. "Why I ever said I would come on this journey I don't know, after my solemn resolve to do no more journeys." He felt that he had let down not only his fiancée but his family: "I broke my word to them all by coming on this journey and I feel I can never look any of them in the face again." But in the men's squalid camp, he added a characteristic note: "Reading *Moby Dick* again. V. G."

Lemon was known to have a hair-trigger temper, and now the suffering and uncertainty seemed to set him off. A few days later, after the trio had made a little more progress southwest in the *Narwhal*, Courtauld wrote in his diary, "Lemon gets more impossible every day. He is continually raising objections to going on. . . . Gino said tonight he was going to have a row if he continued."

By September 19, thirty-six days into their voyage, the men had reached Tingmiarmiut, a small island at the mouth of a deep, skinny fjord, where they found ruins of old Inuit rock-and-sod houses. The last few days of travel, despite the simmering antipathy between Lemon and Watkins, had given the men new hope. "We are well over halfway now," Courtauld claimed in his diary. "If we can only get half a dozen more good travelling days we shall be there."

But the southeast coast was not about to let the men off so easily. Just ahead lay the glacier of Puisortok, which thrust into the sea in its 650-foot-high ice cliff that calved continually and without warning. As recounted in "Interlude: The Cosmos of the East Coast Inuit," generations of natives trying to paddle their umiaks past this dreaded obstacle were forbidden to eat, smoke, or speak until they had safely navigated the gauntlet, and to say the glacier's name out loud was to invite certain disaster.

For the three weary men in the *Narwhal*, Puisortok gave them ten days of "nightmare," in Courtauld's epithet. The weather had turned foul again, and on the first attempt, the outboard motor conked out before they had really gotten started. They ducked into a rocky cove where they made a bad camp. During their stay there, they found the boat had been banged against the rocks so brutally that they had to replace two cracked floor boards and repair the plating on the stern, which had been partly stripped away.

On the second attempt they tried to motor far out to sea, beyond Puisortok's firing range, but pack ice and bergs blocked their way, and again the motor quit. Wrote Lemon, "My patience was now completely exhausted at the miserable performance of the engine, and I was beginning to wish that it would pack up altogether." To no avail, he took apart the carburetor, "a mass of small tubes," in an effort to pinpoint the defect. Only scores of attempts to pull the starter cord as hard as they could, the men taking turns as their arms wore out, could kick the engine into fitful life.

During these dispiriting days, the men's optimism disintegrated. They came to an agreement: if they had not passed Puisortok by October 1, they would winter over—even though, as Lemon observed, "To winter [here] was asking for trouble; there were records of an Eskimo family who had tried to do so having starved to death."

It was during this fraught interval that the tension between Lemon and Watkins exploded, though Courtauld is reticent about the details, recording only that the two men "had their row"—from which he carefully distanced himself. We can imagine the antagonists screaming at each other, even cursing, until Gino reminded Lemon who was the boss of the expedition.

Courtauld's diary between September 23 and 29 is a litany of terse records of failure:

September 23rd. Fine day but had to beach boat to repair leak. . . .
September 24th. Ruined with rain. Spent day repairing boat. . . .
September 25th. Ditto. . . .
September 26th. Ditto.
September 27th. Too much swell.
September 28th. Fine day. Got packed for starting but . . . found the
ice too tight to get on and the fog thick.

In the midst of this roster of disgust, Courtauld pondered longer-range implications. "It is getting sickening to the point of desperation. . . . We have finished the last bit of seal Gino shot the other day but we still have some seagulls," he wrote midway through that week. And on the twenty-eighth: "It really begins to look like having to winter. . . .

The kayaks are getting rotten with all this wet, and if they become useless we have not a hope of being able to feed ourselves in the winter."

On September 29, only two days before the team's self-imposed deadline, they made another try to outflank Puisortok and its calving bergs. Just as they started to feel confident, the motor gave out once more. "None of us said anything while poor Lemon wrestled with the machine," Courtauld reported. "We all realised that if we put back this time it would be for good."

After the umpteenth pull on the cord, the fickle engine "shakily came to life." The men aimed the boat just beyond the field of brash ice spilled from Puisortok and figuratively held their breaths. Whether they observed the Inuit injunction not to eat or speak, none of them recorded; but since they were out of tobacco, there was no temptation to smoke.

At last they motored clear of the nemesis glacier and, with a "good moon," pushed on and on while the engine worked long into the night, choosing a cove for camp only at 2:00 a.m. Exhausted but exhilarated, they rose after only two hours to forge farther onward, but "the engine thought otherwise." They were forced to waste "a brilliant day" on September 30 as Lemon vowed to take the whole motor apart. But now he found that the lug wrenches the team carried were the wrong size for the nuts. He told Gino that it would take him at least a week to manufacture a suitable pair. By then, however, the winter pack might have blocked all access to Prince Christian Sound.

Safety lay tantalizingly near. Only about a hundred miles stretched between the men's camp on the headland of Anoritok and the entrance to Prince Christian Sound, but without the outboard motor, that might as well have been a thousand miles.

During this maddening impasse, Gino looked on but said nothing. Courtauld sensed that he was seeing his leader at the trough of a deeper depression than he had ever before witnessed. "I can see him now in his sealskin boots and brown sweater," he later wrote, "with his hair very untidy and his face very red, looking furiously at the engine, then at the sea which sparkled blue."

While Lemon looked on in disgust, Courtauld tried an adjustment on sheer hunch, pulling loose the silencer and removing the plate that housed it, so that the exhaust would flow straight into the open air. "Immediately the engine went full speed, proving that we had completely cured the trouble." As Lemon, the ace mechanic, later confessed, "Nothing could have been stupider, and I was very annoyed with myself for not having spotted this."

Wildly rejuvenated, the men tore off fifty miles of progress on October 1, even though the *Narwhal* leaked so badly one man had to bail nonstop. The next day they fought a rough sea swell as they rounded a protruding cape, then headed for a fjord they reckoned lay only a day's journey from the mouth of Prince Christian Sound. As they entered the fjord, in Courtauld's words, "we were astonished to see three men rowing in a boat." They were the first other humans the men had laid eyes on since parting from Nicodemudgy's family at Umivik.

The men turned out to be Norwegian hunters dropped off to hunt bears and trap foxes through the winter in this remote fjord. On shore they had built a small hut for a base camp. At once, as Courtauld recounted,

> They took us to their hut and gave us hospitality such as I had never appreciated so much in my life. They gave us a wonderful supper with real bread and potatoes. We rolled up on the floor, feeling we were in the height of luxury to have a solid roof over our heads.

That night the Englishmen enjoyed eight hours of sound sleep, instead of the usual three or four. The Norwegians even shared their tobacco with the newcomers.

One of them, a fellow named Mortensen, proved to be a master carpenter. Turning the *Narwhal* upside down, the men beheld "a dreadful sight: holes everywhere, and all the metal ice-sheathing torn off." It turned out that most of the leaks were through the holes left by the nails that had affixed the sheathing to the boat.

Mortensen went to work with a will, and within hours he had the *Narwhal* more seaworthy that it had been for weeks, with new copper sheathing on the prow. But now their new friends entreated the men to stay and spend the winter with them. They were lonely and homesick, and having found the hunting and trapping on land poorer than they expected, with no means of hunting seals in the water, they were worried about making it through the winter—despite a hefty supply of canned and preserved food brought with them on the ship that had dropped them off.

After three days at the Norwegian camp, however, the *Narwhal* motored on—not before "our Heaven-sent friends" had presented the men with "a cargo of tinned food." It was October 5 when the boat nosed into Prince Christian Sound—just in time, for ice half an inch thick had already formed over much of the strait, ice just barely thin enough for the boat to break through. The next day, near the western exit of the sound, they met Inuit kayakers, and shortly thereafter arrived at the first small settlement on the west coast, called Augpaulagtok, where the Englishmen "received the warm welcome always awarded to strangers by the hospitable Danes."

Two days later the men reached Nanortalik, the first sizeable village on the west coast. They were down to one gallon of boat-fuel when they arrived. Julianehåb was still fifty miles ahead, but here the Open-Boat Journey really ended, as Gino said "God dag" and stepped ashore onto a landing dock that was "swarming like a hive of bees to see the English fools."

Watkins, Lemon, and Courtauld had pulled it off—though just barely. The journey, planned for a month at most, had taken fifty-three days, and the men had narrowly averted mishaps that could have spelled disaster—the struggles with a malfunctioning motor, and with a boat so leaky it might well have sunk, as well as storms so fierce the men could just hold their own in camps in marginal coves.

The joy of arrival on a coast teeming with natives and sturdy houses, with cows and sheep in the fields, with vegetables growing in gardens in the still-green autumn—the satisfaction for Gino of accomplishing the journey that he had looked forward to more than any other in his life, and that should have closed the curtain on the whole grand pageant of the BAARE—was instantly dampened by the news the men received at Nanortalik.

Hampton and Rymill, who should have arrived at Holsteinborg weeks earlier, had failed to show up. There was no word as to their whereabouts. Were they even still alive?

* * *

In Nanortalik the men met Knud Rasmussen. Fifty-two years old that autumn, he was arguably the most famous and accomplished living Arctic explorer (his only rival being the Canadian-born Icelander Vil-hjalmur Stefansson). But if Gino, still only twenty-four, was overawed to meet his hero, it didn't show. According to Courtauld, "[Rasmussen] and Gino at once made friends. It was easy to see how much each admired the other: so similar in their aims though so opposite in their outward characteristics." Rasmussen was indeed impressed by the achievement of the three young Englishmen. As he later told a colleague, "I have seldom seen an expedition with so little equipment and would hardly have believed it possible to make such a journey at that time of year."

Both parties traveled on to Julianehåb together, where they were fêted by the Danish authorities. But now the three Englishmen faced an agonizing decision. A cargo ship in the Julianehåb harbor was ready to set off for Copenhagen. Lemon, who had already strained his good standing with the Royal Signal Corps by pleading for leave to go on the Open-Boat Journey, dared not ask for another extension. But Watkins knew he could not return to Europe with the fate of two of his men undetermined, and Courtauld felt the same imperative to stay. At once the men hitched a ride on the first of a series of local boats that would take them 500 miles north along the coast toward Holsteinborg.

In a letter to his sister, Pam, to be carried back to England by Lemon, Gino could not hide his disappointment. "I just can't tell you how sad I am not to be coming back on this boat," he wrote. "It is wretched, especially as there is absolutely nothing I can do to help Rymill and I am pretty certain he will turn up all right." Courtauld, at least as disappointed, turned his misery into self-abasement, writing Mollie, "It's a bit of typical irony that we should ever have met. If we never had, you would now be happy with someone else and I should be content to follow my fate. . . . For me, it's a wretched state of kicking against the pricks in this revolting Greenland."

More than a month before, on September 4, having dashed across 300 miles of ice cap with record-setting speed, Rymill and Hampton faced only a forty-mile downhill trudge to reach the lowlands. But that last swath of eternal ice loomed now as a Stygian obstacle course. On the eastern edge of the ice cap, almost three weeks before, the men had steered their sledges warily through a labyrinth of crevasses. Now, on the western edge, they encountered an even more fiendish crevasse field. And to top that challenge off, after days of rain and thawing, every level surface had turned into a pool of slush.

Not only did those pools disguise the crevasses, but the men could not avoid wallowing knee-deep and sometimes waist-deep in slush. Soaked through, the men found the dogs almost impossible to manage. At one point, as Rymill led across what looked like a frozen lake, he suddenly broke through and was stranded in three feet of water and slush. To avoid the same trap, Hampton took his own sledge on a detour of several miles to circle the lake, returning to a hypothermic partner. Together they were able to haul the submerged sledge and the frantic dogs to relative safety.

From the start of the traverse the men were burdened with the fore-knowledge (just as Scott's trio had been burdened on their diagonal crossing) that they would have to kill their dogs before reaching the west coast. In Rymill and Hampton's case, that grim necessity was dictated not by the incompatibility of livestock and huskies in their destination village, but by the impossibility of carrying and feeding the dogs through the hundred-mile maze of streams, hills, and cliffs that separated the ice cap from the coast. (Leaving them to starve to death would have been even crueler than shooting them.) Now, after several days of trial by slush, the men decided to abandon one sledge and pack all their belongings and food on the other. They carried out the executions of the ten now "superfluous" dogs one by one with their .22 rifle.

Rymill and Hampton could see the coastal mountains in the distance, but the last few miles of ice cap proved the most hazardous of all. To lighten their loads, they tossed out gear they had once considered essential: the heavy time-set device for gauging longitude, their theodolite, a sextant, two revolvers and a rifle, and their spare tent. After ten days of struggle, they had still not covered the last forty ice cap miles. Slowing them further was the new necessity of carrying the two kayaks (each weighing eighty pounds) ahead, caching them, then returning to bring up the sledge. The ten remaining dogs were so ravenous they ate everything in sight, even tearing off pieces of the canvas tent. Had they gotten near the sealskin kayaks, they would have destroyed them in minutes.

The men would soon have to abandon the second sledge and shoot the rest of the dogs, but until they reached the end of the ice, dogs and sledge were still essential. Otherwise the men would have had to triple-pack the baggage, submitting to the soul-destroying regimen of covering five miles for each mile gained. Meanwhile, Hampton's boots were falling apart. For a week running, he had to fashion soles out of flattened tin cans, replacing them nightly, a job that required "no little skill."

On September 20, seventeen days into the final ice cap gauntlet, they gave up at last on the sledge, which was too heavy for the famished dogs to pull through the deeper and deeper slush pools. With heavy hearts, the men dispatched ten huskies that had seen them through thirty-nine days of all-out toil.

It was not until September 30 that Rymill and Hampton finally escaped the ice cap, which bade them farewell with an 800-foot-long glacial cascade that was worse than anything they had previously negotiated. It had taken them twenty-six days to cover the last forty ice cap miles. Yet now, they were still 100 miles from the coast.

The rest of this desperate trek need only be summarized here. Short of both food and fuel, the men limited themselves to one cooked meal per day. On solid ground, Rymill and Hampton triple-packed: two loads apiece each weighing 100 pounds, followed by the kayak shuttle. Because the map they carried was woefully inaccurate and they had discarded all their navigation gear, they often had to scout ahead without a load to determine a feasible route before returning to lug their baggage.

If only, the men told each other, they could find a stream big enough to kayak, they could pile all their goods into the boats and bomb effortlessly west. On October 10, they found just such a stream. It nearly cost both men their lives.

Paddling ahead, through rapids he had never trained to run, Hampton turned a corner and saw ahead a projecting ice shelf. It was too late to make for the shore. The current pulled him into the shelf, under which the rear end of his kayak momentarily snagged, turning the craft upside down. Rymill turned the corner, saw what had happened to his teammate, and overcorrected, flipping his own kayak. Helpless, he was swept under the ice shelf.

Hampton meanwhile careened upside-down through one rapid after another. He tried twice to execute the Eskimo roll before, with a panicked burst of energy, he dropped his paddle and managed to get out of the kayak. Behind him Rymill had likewise extricated himself and was clinging to a lip of the ice shelf.

Against all odds, both men got out of the water and, after some searching, recovered their kayaks, though Rymill's had to be "fished" out of a crevice under the shelf. Both paddles were gone, and everything the men owned was soaked through. The men pitched a camp immediately as they tried to stop shivering. Somehow their matches had escaped immersion, and after several tries they got a fire going with the highly resinous dwarf birch they found growing all over the hillside.

On October 14, topping a ridge, they saw to their amazement that a fjord stretched in the distance. They had no real idea where they were, but they had reached the coast. In the fjord they saw several seals cavorting, which gave the men hope of hunting the mammals to replenish their nearly empty larder. The next day, after their best sleep in a month, they had just relayed all their belongings to the top of the ridge when they saw, to their even greater amazement, "a good-sized fishing boat."

On the spot they built a fire and fired off several rifle shots, but the boat was too far away to notice their signals. After a fitful night, Rymill and Hampton hiked seven miles along the ridge above the fjord to a point opposite where they had seen the boat. To their relief, the boat lay anchored just below. This time the rifle shots were answered by hearty cheers from the crew.

The fishing boat was manned by four Inuit men, the captain able to speak broken English. Rymill and Hampton learned that the crew had been sent out from Holsteinborg to search for the missing pair, now a month overdue. The men had already been looking fruitlessly for five days. In Chapman's pithy phrase in *Northern Lights*, "In their cheerful Eskimo way the search party had long ago given Rymill and Hampton up for dead, but hoped to find at any rate some remains."

Belowdecks, the captain brewed up tea with plentiful amounts of milk and sugar, and laid out platters of bread, butter, and "tinned delicacies" sent by the governor of Holsteinborg. Rymill and Hampton were mortified by how little they could eat before stomach pains doubled them over.

The fjord where the two men had reached the coast lay sixty miles south of Holsteinborg. On the way there, the captain dawdled, swooping close to shore to show the two survivors each little Inuit settlement on the way. They reached Holsteinborg on October 19, where the whole town turned out to give them a welcome for heroes miraculously delivered from the grave. The governor lent the refugees a house with a servant to wait on them. There they spent a few days "eating enormous meals and and listening to wireless and gramophones."

Only a day or two after their arrival, Watkins and Courtauld came up from the south on the last of their string of local boats. The reunion, of course, was exuberant. A few days later, Gino talked the Danish captain of the *Hans Egede* into making room on his ship for a few more passengers. On November 1—496 days after the *Quest* had delivered the team to Angmagssalik— the last four members of the BAARE to leave Greenland sailed for home.

EPILOGUE

To Slip Betimes Away

Smart lad, to slip betimes away
From fields where glory does not stay. . . .

Now you will not swell the rout
Of lads that wore their honours out,
Runners whom renown outran
And the name died before the man.

—*A. E. Housman,* "To an Athlete Dying Young"

BEFORE LEAVING GREENLAND on board the *Hans Egede*, Gino had managed to get a crucial request to his father, who was in England: "Could you see that a complete suit—bowler hat, shoes, umbrella etc. is sent to the Hotel Angleterre, Copenhagen, as I have got absolutely nothing to wear." His sister, Pam, had arranged to be there on the dock. Knowing the fanfare would be considerable, Gino looked forward to it as a grand entrance.

Courtauld's anticipation was quite the opposite. As desperate as he was to step off the boat into the arms of Mollie, he warned her not to come to Denmark: "There is to be some awful sort of reception in Copenhagen for us, so I should keep clear as these Danes are such fans at formal entertainment." In his diary he added, "There is to be an appalling official welcome. Bands playing, national anthems, Prince Christian, British ambassador. . . . There is a rumour I may have to say something. Heaven forbid! It is going to be an awful show."

The "show" was all that Courtauld feared. The Danish authorities wanted the explorers to arrive dressed like explorers, wearing the soiled and tattered uniforms of their trade. When Gino wired that he had only his old trousers and anorak on board the *Hans Egede*, the authorities wired back, "Anorak sufficient. Good night." But Gino wasn't having it. He wired a newspaper friend, who smuggled the suitcase from his father on board the pilot boat. As Jamie Scott later put it, "And so he was able to step ashore looking as if he had come out of a bandbox rather than the Arctic."

On the dock, decorated for the reception, stood representatives of the Danish government, the British Legation, a swarm of reporters, and a crowd of well-wishers. Pamela Watkins was waiting with Nanny Dennis. After the playing of the Danish national anthem, Gino stepped forward, to the strains of "God Save the King." He spoke modestly and briefly, thanking the Danes for their support of the BAARE from start to finish. The short speech provoked cheers from the public and media alike.

Now a man from the British Legation took the microphone to say, "We must all be particularly grateful to the Danish nation for rescuing Mr. Courtauld." He turned to the rescuee and asked him to say a few words.

The misaccreditation dissolved all of Courtauld's shyness and dread. He seized the microphone and blurted out, "I only want to say that everything the last speaker has told you is entirely wrong. I was not rescued by anyone." According to Scott, the "microphone man" kicked him in the shins.

Bowler, umbrella, and all, Gino tried to escape the reception as soon as he could. But first there was a teatime reception, followed by dinner and a dance. Well after midnight, Scott spirited his friend away from the official fuss and took him to a night club that stayed open till break-fast. Gino must have had a few, for soon, according to Scott, "several tired and unattended ladies within earshot were revived by Gino's con-tinually repeated phrase, 'I'm rich, I'll pay.' "

In the wee hours, Scott and Watkins started talking about the Open-Boat Journey. Full-blown manic by now, "Gino became thor-oughly excited because he was sure he had found the ideal method of summer coastal travel." Apparently forgetting what a close thing the sketchy journey with Lemon and Courtauld had been, or how little hunting seals had really made the difference between survival and star-vation, or how without the whaleboat motor's finally being made to work they would have been stranded, Gino unfolded for Scott a plan for his next expedition that must have been gestating in his brain for months. Next year he would use the same technique to make a circum-navigation of most of the Arctic Circle (all the hard, relatively unknown parts), starting in Labrador, skirting the north coasts of Canada, Alaska, Siberia, and Scandinavia, then terminating at Svalbard. And sometime after that: a complete traverse of Antarctica, Shackleton's last dream.

Suddenly, almost in mid-sentence, Gino broke off this manic fan-tasy. "My God," he said to Scott, "this is a pretty poor way to spend one's first night in civilization." He jumped up from the table, seized an avail-able partner, and poured his energy into dancing.

A couple of days later, Gino returned by boat to England, where his father, on leave from his Davos spa, drove him to London. Not to the home in Onslow Crescent, which further penury had forced his father to sell, but to a hotel on suburban Cromwell Road. According to Scott, "Gino did not seem worried by [the] loss. He was with his family and friends, and that was all that mattered."

Much of the month of November Gino spent preparing his talk for the Royal Geographical Society, scheduled for December 12. To get around London, Gino bought a well-used two-seater automobile that cost him only a few pounds. Scott's account of Gino's handling of this jalopy reads like a vignette out of P. G. Wodehouse. He drove the "almost brakeless vehicle" very fast and with such abandon that passengers, having survived one near-collision, found themselves "gripping the door in anticipation of the next." But Scott had to admit that his friend actually suffered only "one or two minor accidents in his life." The more likely mishap was the car's running out of petrol, for Gino "rarely bought more than a gallon at a time." Apparently he was even more clueless about motor oil. "Does it need oil as well?" he queried a friend. "It's so expensive and I don't know where to put it in. I've never looked under the bonnet."

This portrait of Watkins as a mechanical dunce matches the image of him on the southeast coast staring in anger and frustration at the whaleboat as Lemon and Courtauld tried every trick of the trade to get the outboard motor running. But it's hard to reconcile with the Watkins who got his flying license at twenty and was comfortable piloting a Gypsy Moth over the treacherous fjords of Greenland, or with the man who spent such care and savvy redesigning sledges to meet the quirks of the Labrador outback, or who learned so quickly and so well how to handle an Inuit kayak.

In this case, as so often with Gino, it's not easy to disentangle the poses he struck to amuse his friends from his genuine idiosyncrasies. Did he really need his bowler hat and umbrella to make a comfortable entrée into the social whirl, or did he affect the dandy to offset his real toughness and courage in the wilderness? If so, to what end? One guesses that these contradictions served as camouflage—that for all the loyalty and even love he inspired among his teammates, he needed to be private at the core.

On November 23, Stephen Courtauld, who had donated the vast majority of the funds that made the BAARE possible, hosted a semiformal banquet at the May Fair Hotel for the whole team. The explorers came in tuxedos. The menu was printed in "Eskimo"—as well as the men's grasp of the Inuit tongue could translate such items as *Suprêmes de Sole au vin du Rhin* or *Écrivisses à la Muskoxite*. Above the long dinner table were hung the expedition flag (a polar bear with wings) and the ragged Union Jack from the Ice Cap Station that had been the key to Courtauld's salvation.

Gino was seated next to "a tall fair girl" named Margaret Graham. During his previous social seasons in London, he had flirted, danced, and gone out with any number of women met in similar circumstances, but none of them had become a girlfriend of any duration. Margy, as she called herself, was unlike the others. During their dinner conversation, Gino learned that she too flew airplanes and was widely traveled. Gino invited her to come to the Welsh Harp, a reservoir in northwest London, to watch him and several BAARE teammates demonstrate the Eskimo roll in the kayaks they had brought home. She not only did so, but agreed to race Gino to Marble Arch, she in her Lancia, he in his old two-seater. (Who won the race, Scott does not reveal.)

Gino began "seeing" Margy on a regular basis. She was almost surely a guest at the RGS on December 12, when Gino gave his talk about the Greenland adventure. After Gino's modest presentation, the club's éminences grises spoke with a single mind, as they insisted that the RGS must support whatever grand scheme Gino had next in mind. James Wordie, Gino's mentor, expressed the wish of the whole society "that we may see him setting out again on another expedition as soon as possible." But the most prescient note was sounded by Lauge Koch, the accomplished Danish Greenland veteran, who sensed the fleetingness of deeds such as those the BAARE men had pulled off: "It is only during a short period of one's life that one is able to carry on the work they have carried on. Employ them while they are young.... Time is short."

Into the ears of his receptive auditors at the RGS, Gino poured out his plan for a circumnavigation of the Arctic by kayak (and, one presumes, a supporting motorboat), hunting seals and living off the land as he went, mapping and conducting some kind of ethnography among the various Inuit bands met along the way. But polar veterans such as Wordie and Debenham urged against this grand exploit (for which Gino was willing to allot two or three years), not on the grounds of its impracticality, but because they thought it wasn't as bold a next step as this still youthful genius of exploration ought to take. (As of 2022, nobody has yet completed, or even attempted, an Arctic Circle circumnavigation in the style that Gino so enthusiastically proposed.)

The counsel of the RGS elders pushed Gino instead toward the crossing of Antarctica. As Scott paraphrased their argument:

> They pointed out what Gino had apparently not realised, that the three expeditions he had led had been of rapidly increasing importance and that he must not break the sequence. He must aim at something higher still, and the great geographical problems that remained were in the Antarctic, not the north.

In particular, despite the bold expeditions led by Scott, Shackleton, Amundsen, and Mawson, the most basic facts about the Southern Continent were still unknown. Was it a single land mass, or two? Did the Ross and Weddell ice sheets actually meet, creating a frozen ocean channel right through the heart of the so-called continent? A complete traverse of Antarctica, which Shackleton had intended his ill-fated *Endurance* expedition to accomplish, was the Next Great Problem of terrestrial discovery: 1,500 miles straight through, almost all of it across unknown land.

Gino was quick to come around. Soon he was working day and night on the plan for the traverse. He felt a real urgency, because thanks to the reversal of seasons in the south, he needed to set off for Antarctica only nine months hence.

The plan Gino came up with was characteristically extravagant. In two summers and a winter, his team would not only cross the continent by dog sledge, but it would map some 1,800 miles of coastline surrounding the virtually unexplored Weddell Sea. Gino also proposed airplanes for Antarctica, and maybe even motorized sledges.

Through the first five months of 1932, Gino labored frantically over the Antarctic scheme. None of his advisors seems to have balked at the extravagant ambition of Gino's plan—not even Wordie or Debenham, who had paid their dues in suffering with Scott and Shackleton two decades earlier. The problem was money.

By early 1932, Great Britain was sliding even further into economic depression. The country had gone off the gold standard, and there was scuttlebutt in the newspapers about a total economic collapse. Utterly committed by now to the Antarctic scenario, Gino tried to cut costs any way he could. Debenham wrote a passionate appeal to the public for *The Times*, saying that sooner or later somebody would traverse Antarctica, and British honor was at stake lest the solving of "the last great geographical problem" fall to some other nation. The appeal brought in a few checks of five or ten pounds each.

The BAARE had cost £12,000 to launch, not counting the inevitable overruns incurred during the team's long stay in Greenland. Even whittled down, the Antarctic expedition would cost more—probably much more. Slashing his budget right and left, Gino thought he might just manage to reduce the cost to £13,000.

By late spring of 1932, though, there was no Stephen Courtauld to pour his fortune into Gino's ultimate adventure. On June 20, the expedition committee, every man long-faced, delivered its verdict. Scraping up every resource it might command, the committee could promise a grant of no more than £3,000. Four thousand pounds, perhaps, if true believers far less affluent than Stephen Courtauld were willing to empty their pockets.

The grand plan for traversing the unknown continent was dead.

* * *

As it would turn out, the traverse of Antarctica would have to wait for another two and a half decades. When it was finally accomplished in 1955–58, by a large expedition led by Vivian Fuchs (later Sir Vivian) and Edmund Hillary (already Sir Edmund, after Everest in 1953), dog-sledging played no part in the success. Instead, twelve men rode six tractors across the ice—Sno-Cats, Weasels, and a specially designed Muskeg.

All the while Gino was obsessively planning for the southern continent, his BAARE teammates had dispersed to their "normal" lives. As soon as he could escape the Copenhagen pandemonium, August Courtauld had dashed back to England for his reunion with Mollie Montgomerie. All his fears that she might have tired of waiting for him evaporated. They were married on June 2, 1932, in Southwark Cathedral.

For their honeymoon, he took Mollie across Europe from north to south, then via Sicily to Egypt. In Cairo at first he disdained a visit to the pyramids as too predictably touristy, until Mollie, who could see them from their hotel balcony, insisted. Then the newlyweds boarded a steamer for a slow trip up the Nile to Khartoum. Even that capital of a British colony seemed too full of stuffy diplomats for Courtauld's liking, so he took Mollie to the province of Kordofan, where his old Cambridge pal Charles de Bunsen was a territorial administrator. Courtauld relished not only the absence of tourists, but the Saharan landscape, which he had first encountered in 1927 in West Africa with the vagabond Rodd brothers. And what a tonic, after the Greenland ice cap in winter! As he wrote home from Kordofan, "The country is open bush and sand, dull but pleasant. Mollie has only fallen off [her camel] once. . . . It is rather good to be in a place where it is a known certainty that every day is going to be cloudless and hot."

Back home, he purchased a fifty-foot yacht, the closest one he could find to the ideal yacht he had sketched in his diary while imprisoned in the domed tent. In the summer of 1932 he sailed down the west coast of

Scotland with Mollie. By then she was pregnant, and in the London house in Chelsea Square that August bought, she gave birth to the first of an eventual six children. Despite the Depression, Courtauld was still well enough off to enjoy the life of the idle rich. He threw himself into yachting trips, and soon became one of the best and most daring amateur sailors in England.

On returning from Greenland in the *Gertrud Rask* in August 1931, Freddy Chapman was at loose ends. He was delighted when he learned that Gino had chosen him to write the official account of the expedition. Almost always, the chronicle of a major expedition was the work of its leader: vide Scott, Shackleton, Amundsen, and Mawson. But Gino was too busy planning Antarctica to write the BAARE book, just as he had been too busy planning Greenland to write the Labrador book. Even though he wrote felicitously, authorship was one of the skills that did not come easily to him.

To his aunt and uncle in mid-1932, Chapman anxiously confided, "I think doing the book will take . . . about four months at least. I daresay Watkins will leave it completely to me and let me write it in my own name, *if* the Publishers agree." Of this prospect was born the notion of trying to write for a living. In the same letter, he ventured, "I might as well start to make a name for myself as soon as possible—as a *writer*." But "Freddy Chapman" would not do on the cover or spine of a serious book. Henceforth in print he would be F. Spencer Chapman.

Martin Lindsay also had authorial ambitions, even though he had been appointed to a lieutenancy in the Royal Scots Fusiliers and had led a battalion in Nigeria before joining the Greenland expedition. *Those Greenland Days*, Lindsay's charming account of the BAARE, casts all kinds of lights on the expedition that are different or absent from Chapman's *Northern Lights*. Surprisingly, Lindsay's book appeared a year before Chapman's "official" account. Though this was irregular, no one seems to have minded—least of all Gino.

For one member of the BAARE, the return to England was fraught only with despair. Some months after resuming his regimental duties with the Royal Signal Corps in the autumn of 1931, Percy Lemon tried to commit suicide. The timing is unclear, but the act did not take place before Lemon was able to write the chapter about the Open-Boat Journey for Chapman's book—in which there is no hint of any deep depression on Lemon's part beyond that caused by dangerous storms, miserable bivouacs, and maddening battles with an unreliable outboard motor.

Lemon chose to end his life by swallowing hyposulfite, a chemical used to develop films. He was found unconscious in a London tube station and taken to a hospital. According to one source, upon regaining consciousness he asked for Gino. He recovered sufficiently to be invalided to his parents' house in Brighton, but he remained ill and bedridden for months.

Suicide may so often be inexplicable from the outside, and in Lemon's case, that verdict is especially true. If he ever told anyone why he had tried to end his life, his confidant(s) kept the secret. There are hints that he was devastated to leave his Inuit mistress, Arpika, behind in Greenland. After the suicide attempt, Gino visited Lemon often, both in the hospital and in Brighton. Despite their falling-out at the low point on the Open-Boat Journey, the two men had patched things up. According to Scott, Lemon's devotion to Watkins was so strong that he "would have followed him anywhere."

When Lemon died in the autumn of 1932, Gino was out of the country. Again according to Scott, in his final delirium, "he had been calling Gino's name." Yet Scott too found the prolonged suicide unfathomable. "He was the first apple to fall," he wrote. "He killed himself. There was no reason."

* * *

From the collapse of his Antarctic plans, Gino bounced back immediately. He could not bear to go a whole year without an expedition, and his thoughts immediately turned to Greenland. Despite the achievements of the BAARE, Gino still had unfinished business there. He had turned twenty-five in January: perhaps he was reminded of his flippant remark to Scott that an explorer should have completed his greatest deeds by that very age.

Given the economic doldrums in which the country was mired, and after the futility of his attempts to raise funds for Antarctica, it came as a relief to plan an expedition on a shoestring. Gino thought he could field another Greenland junket—even one lasting a full year—for as little as £1,000. That was still not an inconsequential sum (about $75,000 in 2022 dollars).

Time, however, was now of the essence. Gino had been planning to leave for Antarctica in November, on the brink of the southern spring; but an expedition in the Northern Hemisphere needed to set out by the end of July at the latest. At once Gino wrote to the Danish authorities to reserve places on the *Gertrud Rask*, scheduled to leave for East Greenland in mid-July. Through sheer good luck, in the spring Gino had received a cable from Vilhjalmur Stefansson, who had taken a job as advisor to Pan American Airways. The company was interested in being in the vanguard of any trans-Arctic route linking Europe to North America, and Stefansson had talked its board into offering Watkins £500 for further research on the Greenland link. Now Gino wired back to accept the offer.

In many ways, planning a minimalist expedition appealed to Gino. He had already picked out the locale for a 1932 base camp: not the same fjord in which the BAARE had built its hut, but the inlet, more than 130 miles north of Angmagssalik, which the crew on board the *Quest* had discovered during the Northern Journey in August 1930, and which the team had named Lake Fjord. It seemed to offer ideal conditions both for ice-free boating and for hunting seals.

On his shoestring budget, Gino recognized that he could not afford to charter a ship to carry provisions for a whole year from Angmagssalik to Lake Fjord. Thus the 1932 team would have to rely on Gino's expertise at stalking seals from his kayak to feed themselves throughout the year—which of course had been the core of his plan for the circumnavigation of the Arctic.

The new team, rather than numbering fourteen men, would consist of only four. The first person Gino invited was Jamie Scott. In his son Jeremy's somewhat hostile account, written in 2008, twenty-two years after his father's death,

> Scott had awkwardly refused. He couldn't come, he had been offered a job, he said. It wasn't true and Scott was a poor liar and Gino must have known it wasn't so, but all he said was, "Pity, but keep an eye on Pam, will you."
>
> Scott could not bear returning to the place where he had failed [to rescue Courtauld in April 1931].

In Scott's defense, there were at least two good reasons for his refusal. With Gino's encouragement, he had been dating Gino's sister, Pam. He was in love, and, unlike Gino, was unwilling to go another year without seeing the woman he so cared about. He would get engaged to Pam that autumn, and the two would marry in 1933. At the same time, Scott was busy writing his first book, about the Labrador expedition, which would be published in 1933 as *The Land That God Gave Cain*.

Yet something had indeed gone out of Scott's exploratory spirit after the BAARE. As mentioned previously, he would never go on another expedition, even though he would serve as the stay-at-home secretary for the 1933 British Mount Everest Expedition. Instead, he would return imaginatively and nostalgically to both Labrador and Greenland, in memoir and fiction, in books such as *Portrait of an Ice Cap*, *The Private Life of Polar Exploration*, and *Unknown River*.

In Scott's stead, three BAARE teammates, each of whom had been eager to join Gino in Antarctica, accepted without hesitation his invitation to a pared-down Greenland reprise: Freddy Chapman, Quintin Riley, and John Rymill.

Gino quickly produced a memorandum for Frank Debenham and the *Polar Record*, the widely read journal of the Scott Polar Research Institute, outlining the new expedition's goals. There would be no Ice Cap Station, of course, but Gino promised to keep up weather observations year-round from the coast. The men would map the region of Mont Forel and attempt to climb the peak. Almost impishly, Gino made a virtue of the Spartan conditions the team would have to accept, as he wrote, "We are taking no wood house as there is no room in our motorboats. We shall have to live more or less as Eskimos for a year."

Very tentatively, Gino tacked on another ambition: "Possibly to cross to Godthaab in the spring." This vague suggestion concealed another outlandish plan, one he kept so secret that only one or two people knew about it (among them, Scott). At the end of nine or ten months' stay in Lake Fjord, Gino planned to traverse the ice cap— solo. On paper, such a plan would have looked foolhardy, even suicidal. But if he could not have either Antarctica or the Arctic circumnavigation, Gino needed some new ultimate deed to slake his burning desire to put himself "out there." (The solo traverse would not be accomplished until 1996, when a gutsy Czech explorer, Miroslav Jakes, pulled it off.)

Besides the £500 from Pan Am, Gino got pledges of £200 from the RGS and £100 from *The Times* for exclusive rights to the story. The Air Ministry volunteered the loan of its meteorological instruments. It still didn't add up to £1,000, but Gino was confident he could pull the journey off.

All spring he had been seeing much of Margy Graham. She had joined a Watkins family party at Easter to climb in the Lake District. So often was she a guest at the latest home of the Watkins clan, in rental lodgings in Chester Square, that she might as well have been living there. Sister Pam, brother Tony, Nanny Dennis, and Watkins *père* all liked her. Gino proposed to Margy in early June, and she accepted. His closest friends were delighted, but Margy's mother was not, raising the age-old specter, despite Gino's fame, of "his lack of income and lack of any salaried job to support her daughter."

To placate her, Gino wrote an earnest letter to Mrs. Graham advancing the possibility that he might go to work for Pan Am after the expedition. He also floated prospects of money that might come from a movie the team would make in Greenland about Eskimo life, as well as from a book about the expedition he would write afterwards. Mrs. Graham was so annoyed by the letter that she tore it to pieces and threw it away. (Margy rescued the pieces and taped them back together. The letter is curated today at the Scott Polar Research Institute.)

Despite Mrs. Graham's objections, Gino and Margy formally announced their engagement in June. His friends congratulated him, even as they privately wondered how he could immediately turn around and leave his fiancée waiting at home for twelve months. Tony and Pam apparently had been convinced for years that their headstrong older brother would never get married and "settle down."

The last-minute preparations for Greenland grew frantic, but according to Scott, Gino was "wildly happy" and "careful to make the most of his last weeks in England." At a family party at the May Fair Hotel to celebrate the engagement, Gino and Margy danced past midnight.

Riley and Rymill had gone ahead to Copenhagen. A few days later, Chapman, Gino, and Margy boarded a second ship. "Gino and Margaret Graham lay in the sun on deck reading the chapters of *Northern Lights* as Chapman finished working on them," Scott later wrote. They arrived at Copenhagen on July 13. From the crowd of photographers and fans, Gino and Margy escaped to a cove where they bathed in the warm ocean water. The next day he boarded the *Gertrud Rask*.

Before heading off to Greenland, however, Gino wrote several letters to Margy. They contain perhaps the most intimate thoughts this fundamentally private man ever put to paper. "You are absolutely the only thing in the world that matters to me," he wrote. "It is simply terrible to be going off like this," and "I don't know how I shall get on, as I hate being away from you even for a day. Look after yourself, my angel, and don't forget me!"

* * *

Even before arriving in Greenland, the team of four got unexpected good news on two fronts. Rather than heading straight for Angmagssalik, Captain Tving planned first to visit Scoresby Sound, some six hundred miles farther north. Back in 1922, the great Danish explorer Ejnar Mikkelsen had convinced the government of Greenland to establish a second village on the east coast, in the deepest of all its fjords. Mikkelsen was worried that the growing population of Angmagssalik would outstrip the region's capacity to support its people by hunting. The plan went through, and a sizable group of natives enthusiastically relocated to the new site. But a decade later, many of the Scoresby Sound Inuit were growing disenchanted. The winters were longer and colder there, the hunting was not as bountiful as promised, and some simply missed their old village.

The visit of the *Gertrud Rask* to Scoresby Sound would give a much-needed boost in morale to its denizens, as well as resupplying them with all kinds of European goods. But for Gino's team, it also meant that on the way south to Angmagssalik, the ship would pass by Lake Fjord. Captain Tving promised to try to enter the fjord and land the explorers there, which if he could pull it off would obviate the need to ferry all the expedition gear by motor-boat from Angmagssalik.

In addition, disheartened by the potential collapse of the Scoresby Sound settlement, Mikkelsen had embarked on a personal crusade to build small wooden huts as emergency shelters for hunters along the coast south of the village. Now he promised to try to bring a hut to Lake Fjord and lend it for a year to the Englishmen, which would make for far more comfortable winter quarters than Gino's latest domed tent might afford.

The excitement of the Scoresby Sound natives as the *Gertrud Rask* came into view was redoubled upon discovering four passengers on board who not only knew how to use a kayak but spoke a bit of the Inuit language. As Chapman later wrote, they greeted him, as soon as he had "saluted them in the most colloquial phrases I could remember . . . as if I had been a visitor from another world."

> These natives all had parents, brothers or sisters at Angmagssalik, and as they were rather homesick, questions were rained upon me: Where had I lived at Angmagssalik? Was old Yelmar still alive? Had Salo and Mada any children yet? Who had Beda married?—and so on.

While the *Gertrud Rask* tarried for a few days at Scoresby Sound, the four explorers hiked all over the nearby hills, filmed the first musk oxen they had ever seen, and delighted in the bird life and the profusion of weeds and flowers, including dandelions, "which are a great Eskimo delicacy when eaten with blubber."

On July 29, the *Gertrud Rask* headed south. Two days later, at the mouth of Lake Fjord, Captain Tving instantly saw that it would be impossible to enter the inlet. As disappointed as Gino's team was by the dashing of their hopes for a shortcut, they were buoyed by their return later that day to Angmagssalik. "It was wonderful to meet our old friends again and to hear all the news," Chapman reported. "Old Nicodemudgy was still alive, Arpika and Gertrude were engaged to be married, while Gustari was so grown up that we hardly knew him."

Needless to say, in *Watkins' Last Expedition*, the book he would publish in 1934, Chapman did not mention his shock as he arrived in Angmagssalik again on learning that Gitrude (as most of the accounts spell her name) had given birth to Chapman's son. Nor did he mention his impulsive retort on being told that he must fork over £20 to pay for Hansie's support and education: "Nonsense—he'll go to Sedbergh!" There would be no happy ending. After that brief reunion, Chapman never saw Gitrude again. Six months later he learned that Hansie, rather than matriculating at Chapman's beloved public school in Yorkshire, had died of influenza. Gitrude subsequently married an Inuit man.

Quintin Riley had brought along his own eighteen-foot motorboat, which he named the *Stella Polaris*, for the expedition to use getting their gear and limited provisions up to Lake Fjord and for exploring the intricacies of the inlet. Now he and Rymill took a jaunt in the boat over to the site of the BAARE base camp. It made for a depressing visit, as they found the hut "in a sadly derelict condition, smelling vilely of stale seal-blubber and dirt." The empty place seemed haunted by their memories of men toiling away in the hangar or at the wireless, or cooking hot rum punch and boiled seal meat for teammates just back from arduous treks.

Although he had not been able to steam into Lake Fjord, Captain Tving now did the Englishmen a great favor. After wiring Denmark for permission, he agreed to carry them back up to Lake Fjord before heading east toward Copenhagen. By eliminating the tedious ferrying of baggage in the *Stella Polaris*, Captain Tving would put the expedition three weeks ahead of its program. The *Gertrud Rask* set out on August 8. As pilot, the team brought along an Inuit man named Karali, who had an unusual pedigree and education. Karali was descended from a line of *angakut*, and he had shown an unusual talent for reciting old folktales and illustrating them with drawings. The Danish authorities decided to send him to Godthaab on the west coast where he attended the only college in Greenland. There he was trained as a catechist—a Christian intermediary between the pastor of a village and its native people, charged with teaching the children, healing broken bones, and generally keeping the peace. (As a perk, a catechist was the only Inuit allowed to buy alcohol). The training had transformed him. As Chapman put it, "With his quick sense of humour, his knowledge of the folklore of the tribe, and his quiet intelligent ways he was always a great favourite with any Europeans, though his autocratic ways made him rather unpopular among the Eskimos, who are unused to any sort of discipline."

With Karali's expert guidance, and the warming effect of another ten days of summer, Captain Tving was now able to bring the *Gertrud Rask* well within Lake Fjord. Gino had picked out a "level place" near the head of the inlet that he remembered from two years before as the site for base camp. The ship's crew helped the men unload their supplies. "Quite inadequate, they looked, to support four hungry men for a year, but the memory of the many seals we had seen on our way into the fjord reassured us as we set to work."

As far as they knew, the four men were only the second party of Europeans ever to visit Lake Fjord, their predecessor having been a Danish botanist on a brief stopover in 1911. As the *Gertrud Rask* tooted a farewell horn on August 10, the men set about constructing their home for the coming year. Gino had modified the design that had trapped Courtauld, and this year's domed canvas tent with curved ash ribs had a sleeve door and also a pair of windows made of talc.

If Mikkelsen never arrived with his wooden hut, the men planned to reinforce the tent for winter Inuit-style, by building it over with a sturdy cover of stones and turf. At once the men set feverishly to work: Rymill constructing a bench out of wood from packing crates to serve inside the tent "as a bed by night and a seat by day"; Riley laying out the meteorological apparatus; and Chapman and Watkins building a wind wall and a kennel to house the team's six sledge dogs (four huskies bought at Scoresby Sound and two Alsatians imported from Denmark). The men stacked and inventoried their provisions, a mélange ranging from McDougall's Self Raising Flour to "a case of Cross & Blackwell's potted meats" to something the men called "munch"—a "most delicious" biscuit compounded of oats, sugar, and butter.

It seemed important to start right away on the challenge of "living off the land" to supplement provisions that would not begin to feed the men for a year, so the very first evening Chapman took his rod and reel to fish for salmon in the river flowing into the fjord at its head, while Gino set off in his kayak to look for seals.

One might think that after the grand scale, the bold ambitions, and the journeys ranging from idyllic to desperate of the BAARE, this new Greenland foray would have felt whittled-down, even depressingly tame to the four veterans who had just settled in at Lake Fjord. That doesn't seem to have been the case. In *Watkins' Last Expedition*, Chapman writes about the profound letdown of returning to England in the summer of 1931, to "a cynical, damping world, peopled mainly with business men, whose outlook was so entirely different from our own," and about "the intolerable period" between expeditions. "We had all enjoyed the last expedition so much," he elaborated, "that it seemed to many of us that no future year could ever be so wonderful. To return to the same district and to the same Eskimos whom we knew so well would be joy indeed."

Now, during their first days in Lake Fjord, that joy indeed returned. "It was like the beginning of a marvelous summer holiday," wrote Chapman: "the ideal sort of existence one dreamed of in boyhood."

At Angmagssalik in 1884, Gustav Holm had heard stories from the Inuit about Lake Fjord, which they called Tugtulik, even though none of them had been there. It was a place wreathed in mystical inklings from ancient times. The name alluded to the reindeer that had once thronged the fjord, though they had since disappeared. The natives knew well about the freshwater lake just west of the head of the inlet, which to Gino in the Gypsy Moth in August 1930 had promised a crucial alternative landing site should the fjord itself be icepacked. In that lake, the Angmagssalik people told Holm, the salmon were so large they had to be hunted like seals. They were as large as sharks, and devouring the stomach of a single salmon could sate the appetite of the hungriest husky. And somewhere in the inlet could be found the massive bear trap constructed by the Inuit culture hero Kagsagsik.

Gino returned from his first day's seal hunt empty-handed and a little crestfallen. The problem was not a scarcity of seals—he had seen dozens—but that he was out of practice. He had forgotten, he told his partners, the art of guessing where each kind of seal was most likely to surface after it had dived. But the second day he returned to base camp proudly towing a good-sized seal, with a dozen black guillemots that he had shot stowed in his kayak. That same day Rymill caught eighty salmon in a net the men had stretched across the river just above where it spilled into the fjord. The fish were not actually salmon, the men discovered, but rather a species of char, "but as they looked and tasted exactly like salmon we always called them such." By their second day in Lake Fjord, the men were proving adept at living off the land.

Karali expertly skinned and butchered the seal, and left the bladder intact: the Inuit had long ago learned to use its fat as a source of fuel, a welcome backup now if the men's supply of paraffin for lamp and stove grew short. Until two years before, Karali had been an expert seal hunter himself, but he had given up the pursuit after he had capsized and barely escaped drowning. Now he expressed his concern that Gino was willing to hunt alone. The Inuit almost never did so, not only for safety in case one man capsized, but because it could take two men and two kayaks to tow a heavy seal back to camp.

Compounding the risk was a peculiarity of the fjord. It consisted of two arms. The long main arm stretched some six miles west toward the lake and river inlet, surrounded by rolling hills; the men had built their base camp on the north shore near the head of the inlet. The other arm was a short northern branch, only a mile and a half deep, that headed not in tundra slopes but in the snout of a glacier, which terminated in a vertical ice cliff almost a hundred feet high. Farther inland, this glacier was fed by several tributaries, and in the melting heat of August it looked like a fairly active ice flow. The problem was the seals tended to congregate in that northern branch.

In their rambles among the hills, the men indeed saw no signs of reindeer. The salmon Rymill caught, however, were nowhere near as big as sharks. Nor did the men find the great bear trap that Kagsagsik had built.

On August 12, Chapman and Riley set off in the *Stella Polaris* to return Karali to his home village of Kuamiut, deep inside the long fjord at the mouth of which stood Angmagssalik. The six-day journey turned into a lark, as their Kuamiut hosts treated the Englishmen to a night-long dance with Karali's daughters and their friends—"surprisingly good dancers," Chapman noted, "having practiced repeatedly in the winter evenings."

The men got back to Lake Fjord on August 17. Rymill and Watkins had been busy in their absence, the former laying in a good supply of salmon and starting to map the region, while the latter kayaked back with a killed seal almost every day. The men knew that Ejnar Mikkelsen was making his way down the coast, erecting emergency huts, but since their Spartan preparations had dictated bringing no wireless apparatus, they had no way of contacting the Dane.

Chapman's diary for the next two days details the small delights of the camping life, of the sort that Boy Scouts might have enjoyed in the Cairngorms or the Lake District: "Fried eggs for breakfast. Twelve or so Greenland Redpolls flew over the tent calling." "John makes a kind of bread in the frying-pan." "Played gramophone till midnight: up at 7." "Good old porridge again; you can't beat it." "Heavenly day. John went off to survey."

On August 19, Chapman, Riley, and Watkins spent hours filming the *Stella Polaris* for the movie the team had promised its backers. "Got some good shots, though one camera sticks a bit. Boiled fish for lunch and supper." "I must skin birds tomorrow or they'll go bad. John's fat Alsatian just can't reach his neighbour and they bark at each other all day." "Quintin rearranged the dumps, a good job; Gino never worries about that sort of thing; life's too short."

On the afternoon of the nineteenth, Gino went off seal-hunting, but came back with nothing. In camp that evening, he confessed to an event that disturbed his teammates, though he characteristically downplayed it. Chapman's diary: "He says the other day he all but lost his kayak. He was out on a floe inflating a seal when the glacier calved and with the resulting wave the floe he was on hit the shore and broke in half. He just held on to the end of his kayak."

The program for August 20 was more filming, but when the men rose early to an overcast sky with dull light, they modified their plans. Riley stayed in camp, while Rymill and Chapman took the boat to the point on the south shore where the fjord opens up, about two miles east of camp, Rymill to survey while Chapman indulged his ornithology. Gino set off in his kayak again for the seal-rich northern branch. All three men got off about 8:00 a.m. By early afternoon, Rymill had finished his survey work.

In precise, neutral language, Chapman recorded in his diary what happened next:

> At 2:45 we started to cross the branch with the glacier. At 2:55 I saw a kayak bladder apparently floating, and soon realized it was a complete kayak, without occupant, drifting about. We passed the paddle 150 yards from it. The gloves were pushed in under the fixed seal-skin line on the deck, the gun was missing, the harpoon loaded up but held in position by a strap of ivory beads. We hauled the kayak on board and emptied out most of the water. No sign of Gino. We climbed the mast and looked all around: we stopped the engine and shouted: we cruised up and down—still not a sign.

The implications of their stark discovery were immediately clear to Rymill and Chapman. Yet they could not quite believe that Gino was gone. Not Gino Watkins, who had escaped scores of close calls unscathed, who could survive anything.

In the *Stella Polaris*, the two men motored deeper into the northern branch. They swept back and forth right under the menacing ice cliff of the glacier's terminus. Then, on a small floe only 150 to 200 yards from the glacier, they made a second, utterly puzzling discovery. "We saw his trousers and kayak belt, soaking wet," Chapman wrote. "There was a hollow two inches deep where they had thawed the ice: they must have been there for hours."

Chapman and Rymill tried to recall if they had heard the report of a berg calving off the ice cliff from across the fjord. Independently, they remembered a sharp crashing sound at about 11:00 a.m., but had been surprised to see no wave surging out from the northern branch. Now, up close to the glacier, they could spot no freshly calved berg (the older ones distinguishable by thaw marks along the water line).

Gino's disappearance still seemed inconceivable.

We searched all around for an hour and shouted everywhere. We also searched the hills on each side with glasses. He is wearing his shirt and white anorak, but no boots as he can't get into his kayak with them on.

We decided to go back in case he had already walked home. The accident may have happened several hours ago.

(The present tense in the third sentence betrays Chapman's unconscious refusal to believe that Gino could be dead.)

Back at camp, the two men unloaded the terrible news on Riley. What could have happened? Something must have made Gino's kayak capsize: either a wave from a calving berg, or a bad shot fired too far sideways, or even an attacking seal. In the domed tent, the three men tried to come up with a plausible scenario. Once the kayak had over-turned, they speculated, "In the excitement he lost his paddle and tried to come up with his throw-stick, for that was not on the harpoon." Yet somehow this narrative did not work, for it failed to explain the sodden trousers on the floe. Slowly the men pieced together a different chain of events.

> He may have got out on to a bit of ice to relieve cramp in his legs, to shoot a seal, or to arrange something on his kayak, and then the floe upset of itself or a lump of ice fell off the glacier. This may have carried his kayak off and he may have swum after it, then returned to the floe, for the clothes on the ice were soaked. More probably he was tipped into the water together with the kayak, and then fearing cramp he returned to the floe instead of first recovering his kayak. Then, seeing his kayak drifting off, he must have undressed and swum after it. . . . But he knew we were about and would visit the fjord, so he may have waited an hour or two in the hope that we would find him, then when too late tried to reach his kayak.

Something like this scenario must have unfolded, but no story the men could come up with explained all the enigmatic details. And still they could not accept Gino's disappearance. All three men took the *Stella Polaris* back to the northern branch. At 7:20 p.m. Rymill and Chapman took ice axes and a rope, climbed on shore, and traversed the top of the glacier ("it was prickly ice and not much crevassed"), then hiked overland back to base camp, arriving well after dark.

Finally the three men accepted the inevitable. And only then did Chapman's feelings burst to the surface, to be recorded in disjointed passages in his diary.

> It is too risky this hunting alone. . . . Yet we could not believe that we would never see him again. I felt sure he was dead at this stage but could not grasp it—only a sense of unutterable waste. Gino always dwelt apart somehow, and underneath was as cold and unemotional

as ice: none of us ever fathomed the full and intricate depth of his character. . . . I admire him and feel perfectly happy with him, and I would follow him anywhere. . . .

Gino had his shortcomings but was a very great man. He had complete control over his mind and had the character of a man much older than 25. It is young to leave the world but he has had a lot out of it. He would have hated to die in a bed. . . . Oh why did this happen?

Late at night, on the hike back to base, Chapman had gotten briefly separated from Rymill, "so I had an excuse to go up to the top of a peak as I love to do." There he had a kind of epiphany that briefly assuaged his grief.

It was a heavenly night, with the ageless pinnacle of [Mount] Ingolf silhouetted against the bright yellow and orange of the fading sunset, with hard purple clouds above. A half moon rose higher and higher over the sea, and the stars were almost dimmed by the shaking curtain of aurora. . . . I saw Auriga and Lyra, the Pleiades and Cygnus, all going round the Pole Star just as if nothing had happened—and Gino is dead in the fjord. How shall we carry on without his inspiration? I can't grasp the fulness of the tragedy—he might have done so much, and he is dead.

* * *

The three men spent the next day, August 21, searching again for their lost leader, even though "reason told us there could be no possible hope." An eerie unreality hung over base camp. In Chapman's words,

I think we all felt secretly that Gino could not really be dead. We carried on just as if nothing had happened, mentioning his name quite naturally in conversation, and I think we all kept glancing furtively at the hillsides towards the fjord mouth expecting against the laws of possibility to see Gino's slim figure suddenly appear, quietly apologising for the trouble he had caused us and the untidiness of his appearance.

The evening before his disappearance, Gino had regaled his teammates with the prospect that after this pared-down expedition was concluded, maybe a year later, they would head south together to start the sledging traverse across Antarctica. And oh, yes—on the way they would stop and make the first ascent of Everest.

Now the men realized that the first order of business was to boat to Angmagssalik and get out the word of Gino's death, much as they dreaded that errand. At first they had been uncertain whether to fold up shop and head back to England, or somehow to carry out the original plan for a year's research and discovery. Within hours they had agreed to stick it out, though the men felt that they needed the blessing of Stefansson and Pan Am to stay on. Should they get such a go-ahead, they would obviously need to revise the expedition's logistics, for none of the three was remotely capable of hunting seals from a kayak successfully enough to sustain the team through a year in Lake Fjord. Without debate, they agreed that Rymill should be deemed the new leader, as the oldest of the three and the one in charge of the central mission of surveying and mapping.

On August 22, they battened down base camp, as they "made everything ship-shape in case of a gale in our absence." They also left a note for Mikkelsen, in case he should arrive with his wooden hut while they were gone. On the way to Angmagssalik, they dropped off the six dogs at an Inuit camp in the next fjord south, entrusting them to a native who had already visited them in Lake Fjord.

They spent the first night in Sermilik Fjord, about halfway to Angmagssalik. There they were welcomed by a small community of Inuit, men and women they had already met on their way north in the *Ger-*

trud Rask. As soon as they brought the *Stella Polaris* to shore, they told the acquaintances who greeted them about Gino's disappearance. "The men took it quietly enough," wrote Chapman. "They were very fond of Watkins and were amazed to find such powers of leadership and initiative in so young a man; but they are so accustomed to sudden death, especially in kayak accidents, that they accepted our news with an air of stunned fatality. The women, however, were more demonstrative."

Before reaching Angmagssalik, the three men made another stop at Kuamiut. Karali "was quite overcome to hear of Watkins' death, and blamed himself for letting him hunt alone in his kayak."

At Angmagssalik the *Stella Polaris* was greeted by a "huge crowd." On hearing the news, the whole settlement plunged into lamentation. The governor of the colony immediately ordered all the flags lowered to half-mast. Shortly after arriving, the three men trudged up the hill to the wireless office and sent out their messages. The first was a cable to the RGS in Morse code, asking the society's officers to break the news to Gino's family and to Margy Graham. It was followed by a query to Stefansson asking whether Pan Am would back the continuation of the expedition, and by a long dispatch Chapman had prepared for *The Times.*

The reactions of Pam, Tony, Colonel Watkins, and Nanny Dennis have, not surprisingly, escaped the historical record, but we can imagine their shock and grief. The public outpouring in both England and Denmark was monumental. Both King George V and acting Prime Minister Stanley Baldwin issued statements that transcended the usual homilies of public condolence, Baldwin venturing, "If he had lived he might have ranked . . . among the greatest of polar explorers."

The response in Denmark was even more heartfelt. The Danish Minister in London spoke as if he had lost a friend and mentor: "Why should such a brilliant youth be recalled in the bloom of his years? He stood for me as a representative of the best that there exists in English youth." Knud Rasmussen cabled the RGS from the west coast of Greenland: "Denmark has lost a good friend who always pleaded the cause of the Greenlanders with the greatest understanding."

Hugh Robert Mill, the ancient patron of British polar exploration, composed a moving panegyric:

> I have known, I may say, all the polar explorers of the last half-century, but no one can stand beside young Watkins, save the young Fridtjof Nansen as I met him first on his return from the first crossing of Greenland 44 years ago. . . . Watkins was loved by his comrades, with whom he generously shared the credit for their joint achievements, and he drew the affection of older men in a way I had never seen equaled.

In his dispatch to *The Times*, Chapman let the hints of criticism that leavened the anguish in his diary mutate into a few less tempered remarks. The newspaper leapt upon those cavils. Only ten days after the accident, on September 1, *The Times* published an account subtitled "Warning Disregarded" that quoted Chapman as saying, "As usual, disregarding the expedition's and Eskimos' urgent advice, Watkins hunted alone."

These words infuriated Margy Graham. Several RGS officers tried to soothe her by playing down and reinterpreting the harsh words. A year later, after returning to England, Rymill softened the critique, insisting to the RGS secretary (who forwarded the comments to Margy) that Gino's death "was a lamentable accident, but a pure accident, and due to no error of judgment."

Margy, however, found herself in a painful and ambiguous situation. In *The Watkins Boys*, a follow-up account of the lives of seven of the BAARE men, published in 2010, Simon Courtauld, August's nephew, airs the scuttlebutt from almost eighty years before that he may have gleaned from family gossip.

Over the weeks and months after Gino's death Margy must have wondered whether they would ever have married if he lived. She had his
letters, and all the memories of their intense eight-month relationship, but she knew that neither Gino's family nor his fellow explorers
really believed he would marry—and certainly not in the foreseeable
future. Margy had been very hurt when Chapman told her, just before
the start of the expedition, that Gino did not take his engagement
seriously. [After the 1932–33 expedition] Chapman apologised in a
letter to her, saying "I didn't mean it that way at all."

But in September 1932, Margy took charge of Gino's memorial service,
which was held in the local church near Chester Square, where the Watkins family lived. And much later she donated Gino's letters to her,
along with other memorabilia, to the Scott Polar Research Institute.

After 1932, Margy continued to travel widely on her own. In 1935
she married an air force lieutenant named Humphrey Edwardes-Jones.
After World War II they moved to Sussex. According to Simon Courtauld, the marriage was an unhappy one, and Margy became reclusive.
In later life, she never talked to anyone about her courtship with Gino,
or about the death that ended it.

In Angmagssalik, Chapman, Riley, and Rymill got the go-ahead
from Stefansson and Pan Am to pursue the expedition throughout the
year. At once the men devised a new logistical plan, heavily dependent
on provisions ferried north from Angmagssalik. On their return to
Lake Fjord, they found Mikkelsen's ship anchored in front of their base
camp, having arrived the day before. Not only did the Danish captain
and his men help the team erect a wooden hut, but he gave them quantities of coal and allowed them to buy provisions out of the emergency
stash carried on his ship.

Thus Chapman, Rymill, and Riley managed to carry out a much-
reduced program of exploration and research through the next twelve

months. In the 291 pages of *Watkins' Last Expedition*, Chapman narrates the men's adventures with all the eloquence he had lavished on *Northern Lights*, though the title of one chapter, "Desultory Sledge Journeys," betrays the chastened mood of the long campaign without Gino. The three teammates got through the year without any serious mishaps, though Chapman was laid up for weeks with a bad knee that he had originally injured, ironically, in a crevasse fall during the BAARE.

In March 1933 the men set out on an attempt to climb Mont Forel, but never even got to the mountain, stopped in their tracks by what Chapman swore were "the biggest crevasses I have ever seen in my life." Rymill surveyed away, but the map he later produced was a coast-bound scrap compared to the comprehensive chart he had hoped to compile. With Gino gone, the notion of a concluding traverse of the ice cap to Godthaab was never even discussed. In the summation of one later historian, "Much of the time was spent hunting for food, sometimes with Eskimo assistance, and sailing or sledging to Angmagssalik."

As for the expedition's hopes of contributing anything to Pan Am's plans for a trans-Arctic air route, that fancy was put in perspective in August 1933 when, to the men's complete surprise, a Lockheed Vega (a much sleeker, faster plane than the Gypsy Moth), appeared in the western sky and circled the men's camp, while the pilot waved from his open cockpit. A few days later, in Angmagssalik, the Englishmen met the aviator. It was Colonel Charles Lindbergh, who had crossed the ice cap from Holsteinborg in only five hours, with his wife, Anne Morrow Lindbergh, in the rear seat, in a flight so casual they had shared a picnic en route.

In November the men spent several days constructing a memorial to Gino. Taking two long wooden beams left over from the construction of the Mikkelsen hut, they built a simple cross. Chapman labored carefully over the inscription, carving a simple "RIP" into the vertical beam, and a legend along the crossbeam that read "GINO WATKINS AGED 25 DROWNED AUG. 20 IN THIS FJORD." Just before the ice froze in the inlet, they boated the cross to the point that marks the junction between the two branches of Lake Fjord, where they erected it. The view in all directions was magnificent, dominated by the tower of Mount Ingolf, the highest peak in the region. "Surely Gino himself," Chapman wrote, "would have chosen just such a place to be set apart to his memory."

Sometime during the following decades the deteriorating cross collapsed. But in 1982, to mark the fiftieth anniversary of Gino's death, a group from the British Schools Exploring Society arrived to put up a metal facsimile, bearing almost the same inscription as the one Chapman had carved. It is still visited today by travelers who cherish the memory of England's lost genius of Arctic exploration.

No trace of Gino's body was ever found. It was Riley's opinion, shared with his teammates in the immediate aftermath of the accident, that in all likelihood the body had quickly been consumed by the sharks that shared the fjord with the seals.

Many years later, visitors to Angmagssalik learned a different story about the 1932 tragedy, as told and retold by the natives, until it was transformed into a piece of Inuit folklore. In this version, Gino's cohabitation in the base camp loft with the teenager Tina during the BAARE had resulted in her getting pregnant, then giving birth to a baby who had Gino's fair hair. During his one-week return visit to Angmagssalik in August 1932, a delighted Gino had spent day after day in Tina's hut, playing with the infant. In the near darkness of the hut, however, he had accidentally stepped on the baby, grievously injuring it, and a few days later the child died.

Gino went on to Lake Fjord, the story concluded, but he was so devastated by what he had done that after two weeks in camp he committed suicide by throwing himself into the icy sea.

* * *

It is a curiosity of polar travel that the great leaders have for the most part inspired relatively few teammates to go on to attempt exploits in anything like the same league as their own major achievements. The Norwegian paragons, Nansen and Amundsen, founded no schools of adventurers to complement the attainment of a new Farthest North in 1895 or the South Pole in 1911. Only two of Douglas Mawson's many teammates on the Australasian Antarctic Expedition from 1911 to 1914—Frank Wild and Frank Hurley—went on to forge further deeds in the Southernmost Continent, both as members of Shackleton's *Endurance* expedition of 1914–17.

Two veterans of Scott's and Shackleton's all-out assaults on Ninety Degrees South, Frank Debenham and James Wordie, led expeditions to Svalbard and Greenland in the 1920s and '30s, but nothing on the scale and ambition of their Antarctic campaigns. Debenham and Raymond Priestley, who had survived horrendous ordeals under both Scott and Shackleton, founded the Scott Polar Research Institute, one of the two or three most important polar archives in the world.

The legacy compiled by the men who went to Greenland with Gino Watkins, in contrast, is unmatched in breadth and quality in the annals of modern exploration. No matter that Gino was always younger than his companions, or that his style as a commander was quite the opposite of the rigid top-down leadership exerted by Scott, Amundsen, and Shackleton. The record in the field racked up over more than a decade after Gino's death by what Simon Courtauld calls "the Watkins Boys" is unique. Somehow, each of those young men absorbed from Gino, as if by transfusion, the itch and the nerve to strike out on bold campaigns they led themselves—or performed solo.

In 1933 Lawrence Wager joined the British Mount Everest expedition led by Hugh Rutledge, which included such strong Himalayan veterans as Eric Shipton, Frank Smythe, and Jack Longland. Chosen only at the last minute as a replacement for Noel Odell, who dropped out for business reasons, Wager ended up performing, with his old Cambridge friend Percy Wyn Harris, the most important role on the expedition. The strongest team yet to attempt Everest had the bad luck to run into by far the worst weather yet encountered on the mountain, as storm after storm sent each lead pair into numbed retreat. Finally, on May 30, only days before the summer monsoon would shut the whole range down, Wager and Wyn Harris went into the lead. As they approached the First Step on the northwest ridge, they suddenly discovered an ice axe lying on a slab. It had to have belonged to either Mallory or Irvine, who had disappeared in 1924, and 66 years later, in 1999, its location would form a crucial clue in the discovery of Mallory's body by Conrad Anker.

Wager and Wyn Harris climbed on, without bottled oxygen, turning back only at 28,200 feet, less than a thousand feet short of the summit, the same height at which Edward Norton had given up in 1924. The record for the highest point on earth ever reached without supplemental oxygen would stand for another forty-five years.

When Martin Lindsay joined the BAARE in 1930, he felt that he was the rank novice among far more experienced hands. As he later wrote, "In the *Quest* I appalled my companions by the depth of my ignorance. Although I had joined the Royal Geographical Society some years earlier, under the misapprehension that by so doing I would obtain Sunday tickets for the Zoo, I had only the haziest idea as to what a glacier was."

But the BAARE, and Gino's trust, had transformed him. In 1934 he decided to lead his own expedition to Greenland, feeling "four years older in age but fourteen in experience." The smooth Ivigtut traverse with Scott and Stephenson had whetted his appetite: now he concocted a three-man journey to cross the ice cap from west to east, starting at Jakobshavn and ending at Angmagssalik, but with a crucial detour. The discovery Watkins and D'Aeth had made in their Gypsy Moth of a new range that probably contained the highest peaks in Greenland tantalized Lindsay. By 1934 they were still unexplored and unmapped.

Lindsay's plan was to sledge straight east across the great ice sheet, locate, survey, and map this new range, then head south to Angmagssalik. That would require an unsupported dog-sledge journey of 1,000 miles, longer than any that had ever been accomplished in the Arctic, comparable only to Amundsen's trek to the South Pole and back in 1911.

Lindsay chose two companions for the traverse, neither of whom had any expedition experience. Both were in their mid-twenties; Lindsay himself was still only twenty-eight when the team sailed for Greenland in April 1934. On the ice cap the men ran into the usual problems with crevasses, storms, and sastrugi overturning the sledges. Lindsay grew increasingly anxious about whether the men could even *find* the mysterious range, which must, he was now sure, contain the highest peak in all the Arctic. In late July they at last came in sight of a "great massif" to the south, with one summit clearly outtopping the others. Lindsay rejoiced. "It was the mountain for which we were looking—the highest peak in the Arctic," he later wrote. "We had to call it something when talking about it, so then and there I christened it the Monarch."

But Andrew Croft, an experienced climber, pointed out that such a peak might hide even higher ones behind it. In an agony of uncertainty, the men spent another week getting different views of the range, until finally a series of theodolite readings confirmed the Monarch's preeminence. "I was quite drunk with joy," wrote Lindsay, "and quite incapable of paying attention to detail for the next half-hour."

On their way south the men passed by Forel, long thought to be Greenland's highest. "It seemed a miserable peak," Lindsay sneered, "squat, ugly, and scarcely higher than its neighbours."

At the very end of their trek, Lindsay had the dubious pleasure of sliding down Buggery Bank one last time. Adding up their daily marches, he calculated that the men had actually traveled 1,180 miles, farther even than Amundsen in 1911, across 101 days. For some reason, he waited another twelve years to write a book about the journey, but his slender *Three Got Through* is a narrative every bit as charming as *Those Greenland Days*.

During his solitary ordeal in his ice cap tomb, August Courtauld had vowed never to go on an expedition again, and the miseries and close scrapes of the Open-Boat Journey only reinforced that resolution. The honeymoon with Mollie in 1932 was pure bliss by comparison. Settling down in Chelsea Square to his still affluent life of leisure, he turned his passion to building a family and to sailing. He renamed the "ideal yacht" he had bought the *Duet*, and in it, sometimes with Mollie, but also with his father and with friends, he cruised among the outer isles of Scotland, all around Scandinavia, and past Gibraltar deep into the Mediterranean. He gained a reputation as a highly skilled sailor and navigator with a penchant for pushing the limits. Freddy Chapman, who rode with him on one cruise, commented, "As a seaman he was fearless and undefeated, and frequently frightened his friends."

These were not expeditions; however, in 1934 Courtauld learned that the range Gino had discovered from the air, and that Lindsay had scouted and surveyed on his traverse, had been officially named the Watkins Mountains. Meanwhile, Danish authorities, ignoring Lindsay's "Monarch," had named its highest peak Gunnbjørn Fjeld, after Gunnbjørn Ulfsson, the Icelandic sailor who, blown off course sometime in the late ninth or early tenth century, might have been the first European to see Greenland, a century before Erik the Red. At 12,119 feet, it was indeed the highest peak anywhere north of the Arctic Circle.

Already by 1934 three teams, from Denmark, France, and Italy, had tried to climb the mountain and failed even to reach it. In his typically offhand fashion, Courtauld told a friend early in 1935, "I thought I would get up an expedition to have a stab at the mountains Gino had discovered."

Without much difficulty, Courtauld put together a team of seven Englishmen and three Danes. The aces he counted on for their climbing skill were Lawrence Wager and Jack Longland, only two years off Everest. It was remarkable (but so like Courtauld) that the man he couldn't bear sharing the Ice Cap Station with in 1930 had now become a good friend. (Whether or not Wager ever learned of Courtauld's former antipathy is unclear.) At the last minute, Courtauld invited the wives of the four married men to join the expedition—not for the inland trek and ascent, but to stay on the ship as it docked in Irminger Fjord, awaiting the explorers' return. "Everybody said it was bound to be an appalling failure," he wrote afterward. But "the four girls survived all right and I think they enjoyed it; they certainly made themselves very useful." Courtauld chartered the *Quest*, the ship that had taken the BAARE men to Greenland in 1930. Fighting the worst ice pack in a decade, she finally reached land. Sailing north along the coast, the ship ducked into Lake Fjord to visit the scene of Gino's demise. The big wooden cross that his teammates had erected was scarred with the claw marks of a polar bear that had tried to climb it.

A month after their departure from Aberdeen, the *Quest* finally arrived at Irminger Fjord, and the team at once prepared to cross the hundred miles that lay between the coast and Gunnbjørn Fjeld. Because the approaches were so riven with crevasses, Courtauld chose to manhaul rather than use dogs. Only six of the ten men, including a sole Dane, took part in the inland trek. Three men each hauled two sledges, but sometimes they had to double-haul, and often they parked the sledges while one or two men scouted the route ahead.

Remarkably, after only a week, the team was camped at the foot of the mountain. On August 17 all six men climbed the peak in a thirteen-hour push, with Wager and Longland leading the technical pitches.

So pleased were the men with their clockwork success that they spent another week exploring the range, geologizing and botanizing as they went. On the return to the coast, the conditions had deteriorated, and the men had to navigate around water-filled glacial ditches and *moulins*, the fiendish shafts carved by those streams as they suddenly plunge deep into the ice, but they covered the hundred miles without mishap. Back on board the *Quest*, Courtauld summed up the adventure in his diary, "Very satisfactory. . . . Excellent meal and everyone very happy."

Unlike his BAARE teammates, Courtauld felt no urge to write a book about the expedition he had organized and led so smoothly. But that was enough for one lifetime: the 1935 Greenland adventure was his last expedition.

None of the Watkins Boys took up Gino's mantle more vigorously than John Rymill. In 1934, only twelve months after returning from the expedition on which Watkins had died, he set sail for Antarctica as the leader of an eight-man expedition. Four of his former BAARE companions signed on without hesitation: Wilfred Hampton, whom Rymill made second-in-command, Quintin Riley, Steve Stephenson, and Ted Bingham. Australian-born, Rymill had emigrated with his family to England in 1923, at the age of eighteen. The rare BAARE teammate with no Cambridge connection, Rymill nonetheless filled out his Antarctic team with three more men who had studied at that university.

Rymill freely acknowledged that the original scheme for his southern journey had been sketched out by Gino, who, once it was clear that he could never raise the funds to support a traverse of Antarctica, had briefly toyed with a more limited expedition focused on Graham Land, the long, complex string of islands, mainland, bays, and straits that snakes like a crooked finger a thousand miles north from the bulk of the continent toward the tip of South America. Throughout the rest of the Pan Am expedition (as it is sometimes called), after Gino's death, Rymill had pondered the Graham Land idea, turning it over and over in his mind. Because it was less ambitious than the traverse of the continent, Rymill would later characterize his 1934 scheme as "a modest one as I did not see how I could raise large funds." But the British Graham Land Expedition (BGLE), which would last more than two years, from the late fall of 1934 to the early spring of 1937 (twice as long as the BAARE), would constitute by far the most productive British Antarctic venture since the days of Scott and Shackleton.

Early on, Rymill had invited Freddy Chapman, but after some deliberation he had turned the offer down, taking up the post of a schoolmaster in Yorkshire instead. On the Pan Am expedition Rymill and Chapman had often bickered, and Chapman's diary recorded gripes against Rymill's style of leadership. But the men emerged as friends. More likely, turning down the invitation signaled Chapman's new resolve to go his own way—which he would do during the next decade with spectacular results.

The achievements of the BGLE are too numerous and complex to summarize here. Perhaps its greatest deed was to disprove the confident assertion of Sir Hubert Wilkins (another Aussie), who in pioneering flights over Graham Land had declared that it was an archipelago, a collection of islands stretching from the western edge of the Ronne Ice Shelf all the way north to Elephant Island, where twenty-two men were stranded for four and a half months after the *Endurance* sank in 1916. Rymill's team proved without a shadow of a doubt that Graham Land was a peninsula attached to the Antarctic mainland.

Throughout the two-plus years, the men survived serious near-accidents, including the obligatory plunges into crevasses. But no member got seriously hurt. Before the team was done, it had surveyed a thousand miles of previously unmapped (and virtually unknown) coast. Many years later Colin Bertram, one of the members, swore that it was Rymill's "superb leadership and technical excellence on the broken ice that saved us all. Had he failed all would have been lost."

On his return to England in 1937, Rymill received several citations and honors, including the Founder's Medal from the RGS and a rare second bar added to the British Service Polar Medal he had earned after the BAARE. For the book he wrote about the expedition, he perhaps cheekily chose the title *Southern Lights*, to echo Chapman's *Northern Lights*, the BAARE account. In contrast to the lively prose that Chapman and Scott so effortlessly commanded in their books, Rymill's style was workmanlike, sometimes even wooden and always modest. But the man suffered from lifelong dyslexia, and to produce a coherent 328-page book at all was a great achievement.

Freddy Chapman's Yorkshire schoolmastership did not last long. On summer holiday with his students, he met Marco Pallis, a scholar and explorer with considerable experience in the Himalayas. Planning an expedition for the following year, he invited Chapman only days after meeting him, and Chapman promptly accepted.

On the BAARE, Chapman had been the second-most-talented climber, after Wager. But while Wager had been accepted for the 1933 Everest expedition, Chapman's application had been turned down. Unfazed, the man came into his own in the Himalayas in 1937. With a single partner, the Sherpa Pasang Dawa Lama, he attempted a major unclimbed peak in the Kangchenjunga massif called Chomolhari, a stunning 24,035-foot pyramid of rock and snow that had defeated other, larger teams. Chapman and Pasang pulled off the ascent without incident. But the descent, which should have taken one day, devolved into a five-day ordeal, during which Chapman arrested a 300-foot roped fall by both men and cut steps in vertical ice to escape a crevasse into which he'd fallen.

Pasang would go on to become a legendary climber in his own right. In 1939, he got within 750 feet of the summit of K2 with Fritz Wiessner, and in 1954, he joined two Austrians to make the first ascent of Cho Oyu, the sixth-highest mountain in the world.

During World War II, nearly all the Watkins Boys—now men in their mid- to late thirties—served in the various branches of the military. Some were later honored for their heroism or for being wounded in battle, including Martin Lindsay, who led troops in pivotal campaigns in Norway and Italy. None of them, however, waged a wartime campaign quite so perilous and brilliant as that of Freddy Chapman, who agreed to go underground behind the Japanese lines in Malaya, where he fought a three-year guerrilla campaign, most of it as a solo saboteur, against the invaders. He was captured by Chinese bandits and by Japanese troops, but escaped from both. At one point he lay for seventeen days in a malarial coma. His feat is regarded as one of the great survival and resistance stories of World War II, and the book he wrote about it, *The Jungle Is Neutral*, a bestseller in its day, is now regarded as an enduring classic.

All of the BAARE men (except, of course, Percy Lemon, dead at thirty-four after his suicide attempt) eventually married. In various permutations, they kept up for the rest of their lives friendships first forged in Greenland. Their marriages ran the predictable gamut from successful to troubled, and a fair portion of them ended in divorce. Jamie Scott, nudged by Gino, had courted and married Gino's sister, Pam, but according to their son Jeremy, the marriage was doomed from the start. Twenty-two years after his father's death, Jeremy Scott coldly assessed the man's marriage and career:

> It was a disastrous mistake for both of them for they had nothing whatever in common except, in time, three sons, the eldest of them myself. During World War II Scott trained special forces in irregular warfare, survival and silent killing. After that . . . he wrote books. In the course of his life he published thirty-five, two of them bestsellers. Yet nothing gave him happiness. He hated family life, ties, responsibilities, job, routine. . . . In the Arctic he had found the only existence that delivered.

The Scotts divorced in 1958, and soon thereafter Jamie married an Italian woman with whom he seemed to find lasting harmony.

According to Simon Courtauld, after Gino's death the ever-loyal Nanny Dennis moved in with the Scotts, but "She was paid a pittance and died of malnutrition and cancer in her eighties, poorly treated by Pam towards the end."

The happiest of marriages was surely the union between August Courtauld and Mollie Montgomerie, even though (or perhaps because) he never took a real job, spending one holiday after another sailing the *Duet* to far-off places across fickle seas. Yet that most self-effacing of teammates, whose sangfroid during his unfathomable entombment in the ice cap tent derived from his faith in God and the conviction that his teammates would somehow liberate him, was destined for a bad ending. In 1953, at the age of forty-nine, he was stricken with multiple sclerosis. His last six years were measured out by the deterioration of one function after another, as he gradually lost his eyesight and control of his body, until finally, as a terminal blow, the disease robbed him of his mind.

During the first years, unwilling to accept what the doctors told him was inevitable, Courtauld continued to plan voyages, and when he could no longer skipper the *Duet*, he hired a professional sailor to take the helm, while he rode as a passenger confined to his chair, carried up to the deck and perched in the cockpit to watch the inexhaustible world slide by. After Courtauld's death in 1959, Mollie lived on to the age of 101, becoming the last living person who had known the Watkins Boys in their prime.

There were other bad endings. At least two of the men became serious alcoholics in their later years—Quintin Riley, whose powers deafness and a brain hemorrhage seriously curtailed, and (if Jeremy can be trusted) Jamie Scott, despite the steady output of new books until near the end. Two men died in auto accidents: Riley, after another car ran a stop sign and into his path; and John Rymill, in Australia, when he fell asleep at the wheel and his auto left the highway and hit a tree. Rymill lingered in a coma for six months before dying.

Freddy Chapman, the toughest of all of Gino's boys, was partly incapacitated in his sixties by a recurring bowel disorder, chronic back troubles, and frequent headaches. Unwilling to accept the reduction of his powers, and assailed by constant pain, he shot himself in his study in 1971, at the age of sixty-four, leaving a note for his wife: "I don't want you to have to nurse an invalid for the rest of my life."

Among the BAARE principals, perhaps only Martin Lindsay enjoyed a serene and honored old age. After the war he was knighted by

Queen Elizabeth, who also bestowed on him the hereditary title of bar-onet. Lindsay served many terms in Parliament, and in his latter years was much consulted for his wisdom on both politics and business. He died in 1981, at the age of seventy-five. Twenty-six years later Lindsay was revealed as the author of the secret memorandum excoriating the early British campaign against the Nazis in Norway, which toppled Prime Minister Neville Chamberlain and led to the installation of Winston Churchill at the onset of the war.

* * *

"He was a man, take him for all in all, / I shall not look upon his like again." So Hamlet muses about his father, the dead king of Denmark. And such was Gino Watkins.

His most startling talent lay in his precocity, as the leader of four Arctic expeditions by the age of twenty-five. A comparison with his most celebrated peers in polar exploration underscores that unique-ness. At the age of twenty-five, among the members of the pantheon, only Roald Amundsen had even been to the Arctic or Antarctic, and that only as second mate on the ill-starred *Belgica* expedition, the first to winter over in Antarctica, thanks to the incompetence of its commander.

Ernest Shackleton was forty as he set out on the *Endurance* expedi-tion that would win him his lasting fame. Robert F. Scott was forty-two when he headed for the South Pole in his race with Amundsen. Robert Peary was fifty-two, his toes long gone to frostbite, worn-out and embit-tered on his eighth attempt to reach the North Pole, when he "nailed the Stars and Stripes" to that imaginary post, though still some hundred miles short of it. Even Fridtjof Nansen, another precocious leader, was twenty-seven when he completed his first great Arctic venture with the pioneer crossing of Greenland.

Another measure of Watkins's excellence was that on his four expe-ditions, he never lost a man—except himself. Scott, of course, perished with his four brave teammates after reaching the South Pole a month after Amundsen, too depleted to manage the long trek north back to base camp. Douglas Mawson's unimaginable survival feat on the Australasian Antarctic Expedition was necessitated by the shocking deaths of his two companions, Belgrave Ninnis and Xavier Mertz. Shackleton's excellence as a leader is often buttressed by the fact that he saved every last man of the twenty-eight who were stranded when the *Endurance* got frozen into the pack, then sank. But that claim to excellence ignores the grim fact that three members of his supporting team on the other side of Antarctica, charged with laying depots for the home stretch of the traverse that never happened, died in the effort (a tragedy that Shackleton himself never downplayed). Even Knud Rasmussen lost two companions on the Second Thule Expedition, a disastrous traverse of Greenland along the northern fringe between the ice cap and the coast.

The two great Norwegian explorers, Amundsen and Nansen, deserve the highest praise for leading two perilous polar expeditions each without losing a man.

Much of the enigma of Gino Watkins lurks in his leadership style, a quirky blend of laissez-faire and demonstration by example. Even his teammates found it hard to pinpoint that style, instead trying to illustrate it by citing, as Lindsay did, the apparently offhand suggestion—"I say, Martin, do you mind going up with Jamie to relieve August?"—as the epitome of his commandership. As the youngest man among the fourteen members of the BAARE, Gino could hardly bark out orders à la Scott or Shackleton. Yet none of the four expeditions could be called a true campaign by consensus. Without Gino, all four would have fallen apart—or never gotten off the ground.

Some of his more independent-minded teammates (notably Lindsay and Chapman) had their private quarrels with Watkins. Yet none of them ever really struck out on his own, or attempted a deed that was at odds with Gino's main plan. He had an unworldly ability to stay calm during the worst emergencies, to sleep like a baby in the most miserable, storm-lashed camps. One way or another he effortlessly won the loyalty of all his teammates. So many of them later uttered that ultimate pledge, "I would follow him anywhere."

And yet, what a parcel of contradictions was Gino Watkins! He could act like a teenager, dancing till three o'clock in the morning with Inuit women or playing the gramophone till the needle wore out. Back in England, he dressed like a well-tailored fop, down to the bowler hat and the rolled-up umbrella on the sunniest days. When he wanted to edify his teammates with visions of future adventures, he could spin the most outrageous riffs. If you met a fellow in a pub who suddenly sketched out a plan to knock off Everest on the way to traversing Antarctica, you'd be inclined to dismiss him as a poseur or a drunken fool. But balancing these riffs, and the illusion Gino gave off of a habitual carelessness, was the meticulous and innovative planning he poured into every stage of a journey, from redesigning sledges to devising the ideal cold-weather diet. Unlike Scott and Shackleton, or any of a whole line of British Arctic explorers in the nineteenth century, Gino was eager to learn everything he could from natives who had taken centuries to perfect the crafts of dog-sledging or hunting from a kayak. He was the least squeamish of all his men when it came to eating Inuit food.

One senses that at the heart of Gino's character was a deep need to remain private, even unknowable. Chapman captured that quality in the remark in his diary on the day Gino disappeared: that "Gino always dwelt apart somehow, and underneath was as cold and unemotional as ice." The poses, the affectations, the outlandish riffs about future exploits all served as camouflage.

Some commentators have compared Watkins to T. E. Lawrence— Lawrence of Arabia—in this respect: the genius for command, combined with a severe need for privacy. (In *Goodbye to All That*, Robert Graves, who knew Lawrence well, noted with surprise that the man abhorred being touched, even in the most casual way.) If, as Simon Courtauld insists, Gino's family and friends all believed he would never go through with the marriage to Margy Graham, what end did it serve so to delude her? Or had she, as his letters to her seem to hint, broken through his shield and unlocked the gates of intimacy?

However flighty or impulsive Gino might be, he was never gratuitously cruel. Without attempting the absurd task of putting Watkins on the psychoanalytic couch, one might wonder what the impact on him of his mother's suicide must have been, and how that linked up with his unquestioning love for Nanny Dennis, who even when he was twenty-four would still bring him his Ovaltine in bed in the morning.

It is a commonplace today, even in Britain, and even among the polar cognoscenti, to refer to Gino as "a forgotten hero." His early death is inevitably cited as the primary reason for that neglect. But another factor is the inordinate, even absurd emphasis on the North and South poles as the only exploratory goals that mattered. (A kindred fixation, lasting through the present day, is the public obsession with Everest.) Even Nansen, on his brilliant 1894–97 expedition in the *Fram*, got caught up in the frenzy to be first to reach the North Pole. A good argument can be made that Amundsen's solving of the Northwest Passage from 1903 to 1906 was a greater deed than his attainment of the South Pole in 1911; but if his laurels rested mainly on the brilliant voyage of the *Gjøa*, his fame would shine far dimmer today. The great Adolf Erik Nordenskiöld, author of the Northeast Passage, is indeed a forgotten hero today.

Watkins, like Douglas Mawson, was uninterested in poles, and the fame of both men has suffered as a result. Yet one cannot help returning to Gino's death at twenty-five as an accident that robbed exploration of all kinds of future triumphs. His early demise invokes Santayana's touchstone for tragedy, that it makes us sigh and say to ourselves, "Oh, what might have been!"

Stanley Baldwin, striving for a valedictory note, could not help lamenting, "If he had lived he might have ranked . . . among the greatest of polar explorers." But after Franz Schubert died in Vienna at the untimely age of thirty-one, a French critic summed up his achievement with the judgment (absurd to today's ears) that had he lived longer, Schubert might have become "the equal of Carl Maria von Weber." Instead we cherish the astounding outpouring of masterpieces Schubert did compose, especially during his last years, when he was dying of syphilis.

So we should cherish Gino Watkins, not for what he might have accomplished in a longer life, but for what he did and who he was through the first half of his twenty-sixth year. For all his contradictions, for all his inner privacy, he blazed a meteoric trail across the Arctic and into the hearts of his teammates that fades only with the pitiless erosion of time.

We shall not look upon his like again.

ACKNOWLEDGMENTS

As I began researching Gino Watkins in the spring of 2020, the coronavirus pandemic had recently shut down libraries and archives all over the world. Fortunately (or so I thought), most of the primary materials I hoped to consult were housed in two facilities: the Scott Polar Research Institute in Cambridge, UK, and the Royal Geographical Society in London. Normally I would have planned a trip to England to start digging, but already borders worldwide were closed tight to foreigners. In addition, the Harvard libraries, on whose nearly limitless resources I've depended for nearly all my books, were shut as firmly as their British counterparts (though an online site called Hathi Trust, available to Harvard alumni, allowed me to read important books published before 1925).

I fretted away for several months, hoping the situation would ease. A Harvard freshman, Taylor Larson, was able to snag three rare and important books for me via interlibrary loan, as well as to open several journal articles. Meanwhile I read everything I could get my hands on about Gino and the companions on his four expeditions. I spent a small fortune on out-of-print tomes through the wonderful Bookfinder.com site. In those expedition accounts and memoirs there was such a wealth of human detail, remembered conversations, and quotes from diaries I had no hope of reading in 2020 and 2021 that I was soon awash in a sea of rich material. Details the men themselves were too squeamish or "proper" to commit to print emerged in latter-day biographies and in the uncensored revelations of Simon Courtauld and Jeremy Scott, younger relatives of two of the principal actors in the Greenland drama. As the year wore on with no sign that archives and libraries were likely to open soon, I decided to plunge ahead and start writing anyway. As of June 2021, the research dilemma remains fundamentally unchanged. I'm glad I went ahead.

Despite the success of Gino's first two expeditions, to Edgeøya (in Svalbard) and Labrador, it was obvious that my book should focus on Greenland, the arena of his greatest deeds and harshest setbacks. I'd been to Edgeøya, but I'd never been to Greenland. Ideally, a journey to its east coast, centered on the village of Angmagssalik, would have deepened my grasp of that magnificent Arctic island. But in 2020 Greenland was shut as tight as England, and in any event, my fragile health after five years of cancer would have made that trip risky and even futile. (On my last trip abroad, to France in 2019, I'd collapsed and been rushed to the nearest hospital. Sketchy as the medical facilities were in Carcassonne, I suspected they were several notches above the Angmagssalik ER.) Still, I'll regret for the rest of my life never having seen or explored Greenland.

My longtime friend and fellow author Larry Millman, who's been to Greenland many times, as well as to many other Arctic outbacks, and has written brilliantly about all of it, helped me out with both insights and encouragement. Another longtime friend, Irere Owsley, who's visited Greenland twice, gave (not lent) me three expensive, rare, and useful books about the island's history and the great ice cap.

As usual, my wife, Sharon, was my first reader, commenting on each chapter as soon as I finished writing it. Her canny strictures saved me from many a minor embarrassment, and she pointed me in directions I wouldn't otherwise have pondered.

Three of my best friends, Matt Hale, Ed Ward, and Michael Wejchert, also read my manuscript chapter by chapter. Matt was his perspicacious best at catching my careless oversimplifications, and he pleaded with me to cut some slack when I played armchair critic to some of Gino's team. Ed and Michael responded by comparing the ordeals and triumphs of the men in Greenland with their own long pedigrees of mountaineering breakthroughs. I knew better than to confuse my buddies' wholehearted enthusiasm for Gino's story with the responses of potential readers, but their page-turning zest gave me heart when I was wondering where to go next. Jon Krakauer took a longer view, reading big chunks of the manuscript at some remove from the ink drying on the page, and his sharp critique guided many of my revisions.

Throughout the year during which I wrote about Gino Watkins, I was under the care of an oncologist at Dana-Farber Cancer Institute named Dr. Jochen Lorch. In fact, Dr. Lorch had treated me since September 2017, and his very first act was to substitute a pair of new (and not yet FDA-approved) immunotherapy drugs in place of pembrolizumab, which seemed to have given up the fight against my latest metastasized adrenal tumors. That canny decision, I now believe, has kept me alive for the last three and a half years.

Since then, I've seen Dr. Lorch almost every other Thursday for my regular infusions and to check out all my other ailments. I quickly realized that the oncologist, German-born from Stuttgart, is a man of high learning, sharp wit, and generous curiosity. Our meetings over the years have digressed into chats about Karl May's cowboys and Indians in the Black Forest; Thomas Mann's *Buddenbrooks* versus Graham Greene's *The End of the Affair*; the glories of Romanesque architecture in the Bourgogne (as a lad, he trampled grapes in the vineyards of Fleurie); Goethe tapping out hexameters on the shoulder blade of his latest lover; even the joys of camping and hiking, so curtailed for me now.

I dread everything about hospitals, but I soon came to look forward keenly to my séances with the good Dr. Lorch. Then, just after I finished writing this book, he broke the news to Sharon and me that he was moving to a better position in a hospital 1,000 miles from Boston. I felt devastated. I'd had two very good oncologists before Dr. Lorch, and no doubt my next will likewise be excellent. But to find a physician who can save your life while sharing with you the passions and stories that would normally render you longtime friends—that's as rare as a nightingale in New England. May he flourish in his new calling, while I survive a little longer in my old one.

As he did for my previous book, Charles Lecompte devised and drew the maps for *Into the Great Emptiness*. In *Bears Ears: A Human History* he needed to craft only a single comprehensive chart, but Gino Watkins stretched him to five separate maps. These route guides are as elegantly drawn as they are historically accurate, and I feel fortunate to be able to include them. Charles also handled the sometimes Byzantine process of securing permissions for the photos reprinted here—no easy job, with the host institutions closed to visitors—as well as negotiating with publishers for rights to quote liberally from memoirs published almost ninety years ago.

Once again, I was lucky enough to have Star Lawrence as my editor at W. W. Norton and Company. For the last nine years Star has been both my loyal champion in the face of sometimes hesitant colleagues and a sharp arbiter of where I go wrong. This time around, we hit a major snag. There's no getting around the fact that I got carried away writing about Gino. When the word-dust settled, I discovered to my horror that I'd committed some 70,000 more of the precious little gremlins to paper than I'd promised in my proposal. Star was justifiably miffed. (I'd only committed this Thomas Wolfe–style excess once before, way back in 1986.) In a feverish month of slashing—so much harder for me than writing—I pared the narrative back to its (I hope) reasonable length. And much as I hated to lose stories I was in love with, I had to admit the book was better for the revision. Star could have told me to take a long hike: instead he seemed to rediscover why he wanted my books in print in the first place. May I here salute both his wisdom and his loyalty.

At Norton, Nneoma Amadi-obi handled all the fussy details of editing and permissions with her usual acumen, and Erin Lovett once again went to bat for me as only a publicist who loves her work can.

What new can I say about my agent, Stuart Krichevsky? I've run out of superlatives. This is our eighteenth book together, spanning twenty-two years. Way back in 1999, when I finally decided to search for a new agent, after a decade of collaboration with a kind but hopelessly old-fashioned fellow, I interviewed with the three New York agents known as the leading champions of writers with a zest for adventure. I was wavering between the other two until I met Stuart. It took me three seconds to decide.

Over all these years, Stuart has not only pushed me in directions that blended my instincts with marketability, but has warned me sternly away from what I thought were great book ideas, whose bubbles he burst with single stabs of his critical needle. No magazine editor among the scores I've worked with (with the sole exception of John Rasmus) has possessed such uncanny judgment. (I keep urging Stuart to write a memoir, and he keeps demurring. It's not shyness: it's his rock-steady faith that his calling is to cultivate talent, not tell tales out of school.)

If Stuart can forgive me for quoting one of his e-mails, I will do so for the sake of underlining his fundamental humanity. On May 28, 2021, I wrote him, "Tomorrow is my 78th birthday. Rain most of the day, high of 51 degrees. I shall celebrate by cutting 10,000 more words out of Gino's soul." Almost every writer, long before the age of seventy-eight, fears the void that turns a solid "midlist" author into an unpublishable wretch. Even without cancer.

But Stuart wrote back, "That is incredible! It has been wonderful to have you with us for what now feels like a five-year bonus period (ever the agent) when we certainly haven't taken your presence for granted. And a few good books in there, too. Happy birthday, and sorry, Gino."

The Stuart Krichevsky Literary Agency is staffed by colleagues skilled enough to launch out on their own, which Stuart encourages. (Other agents see only betrayal.) Two of them who've helped me immensely over the years are Laura Usselman, who arranges everything to do with foreign rights, and who swung a great multi-audiobook deal for me; and Aemilia Phillips, who makes sure we scriveners get paid—and promptly, too. One need only remember other adjutants who made it as hard as possible for an author to cash his checks or expand his audience to realize how smoothly Laura and Aemilia operate.

The year and a half during which I researched and wrote *Into the Great Emptiness* coincided, as mentioned above, with the worldwide Covid shutdown. The isolation and sense of loss the pandemic fostered was very hard not only on writers, but on couples. During almost all that time, Sharon and I were alone together. Travel was limited to short trips to nearby Airbnbs, where the isolation seemed little different from home. Restaurants were closed, and even shopping at a market could seem a risky act.

In the acknowledgments to my previous books, I've praised Sharon to the skies for her unwavering commitment to keeping me alive, and from those vignettes folks who've never met her have constructed the image of her as a "saint" (a designation she disowns.). But it's high time I recognize here what Sharon gave up, when she resigned from her very successful career as a psychoanalyst in private practice. In 2006 she became the first woman to be elected president of the Massachusetts Institute for Psychoanalysis, one of the highest achievements in her profession.

Two or three times during the last year and a half I've found Sharon in tears, after chatting with a colleague who was still active in the field. Only then did I really glimpse the magnitude of her loss. Yes, she deserves all the credit I gave her for taking care of me through my endless illnesses (including four more hospitalizations between October 2020 and June 2021); but to thank her for that alone is to wallow in my own self-centeredness.

The last six years since I "got" cancer have been tremendously stressful for Sharon, but she's never wavered in her determination to keep me alive. But it's high time to acknowledge her own accomplishments as a professional, and the sacrifice entailed in giving up her career to take care of a difficult husband.

If it was Sharon who had come down with cancer, would I have given up writing to take care of her? I like to think I could say yes, but part of me cannot go beyond "maybe." What I owe her for her six years of sacrifice (and counting) is not only devotion and love, but a full appreciation of the heroism it took for her to make that choice—and having chosen, never to look back.

NOTES ON SOURCES

NB: PLEASE CONSULT the bibliography for full citations of sources.

Prologue: The Man on the Ice Cap

The main sources are Scott's biography of Gino Watkins, Chapman's *Northern Lights*, and Lindsay's *Those Greenland Days*. Excerpts from Courtauld's diary come from Wollaston's biography, *The Man on the Ice Cap*; and from Scott's *Portrait of an Ice Cap*.

Chapter One: "He Never Discussed Anything Seriously"

Virtually all the details of Gino's childhood, adolescence, and first year at Cambridge come from Scott's biography. The routine of student misery at Lancing College is vividly captured in Waugh's *A Little Learning*.

Chapter Two: Edgeøya

The discovery and early history of Svalbard are admirably recounted in Conway's *No Man's Land*. Virtually all the narrative of Gino's expedition to Edgeøya derives from Scott's biography; some additional details come from Gino's report in *The Geographical Journal*, "The Cambridge Expedition to Edge Island." For Gino's doings during the year after

Edgeøya, Scott's bio is again the source. The shocking suicide of Gino's mother is most fully reported in Jeremy Scott's *Dancing on Ice.*

Chapter Three: The Land That God Gave Cain

Virtually all of the Labrador chapter summarizes the narrative in Scott's masterly expedition account (sadly forgotten today), *The Land That God Gave Cain.* Additional sidelights come from Scott's bio. A succinct account of the Quebec-Labrador border dispute appears in Alec McEwen's online essay, "Labrador Boundary Dispute."

Chapter Four: Gino at Home

Once again, all roads lead to Scott: in this case, to the biography. Gino's talk to the Royal Geographic Society about Labrador was published in *The Geographical Journal* as "River Exploration in Labrador by Canoe and Dog Sledge." Some of the details of the early connection between Courtauld and Watkins appear in Wollaston's bio of Courtauld. The anecdotes hinting at Gino's bisexual or ambisexual nature come from Jeremy Scott's *Dancing on Ice* and Simon Courtauld's *The Watkins Boys.*

Chapter Five: Base Camp

With the voyage of the *Quest* to Greenland, the bulk of the source material shifts to Chapman's *Northern Lights,* with liberal contributions from Lindsay's *Those Greenland Days.* The details of the Danish protocol guarding the Inuit against sexually transmitted diseases and unwanted pregnancies appear in Barker's biography of Chapman.

The European "discovery" of Greenland and initial colonization by Erik the Red are well told in Gad's *The History of Greenland* (volume I). An excellent summary of the current archaeological and ethnographic understanding of Inuit, proto-Inuit, and pre-Inuit peoples can be found in "The Prehistory of Greenland," on the website of the National Museum of Denmark (https://natmus.dk/organisation/forskning

-samling-og-bevaring/nyere-tid-og-verdens-kulturer/etnografisk
-samling/arktisk-forskning/prehistory-of-greenland/). Egede's recolo-
nization of Greenland is covered in volume II of Gad's *The History of
Greenland*.

Graah's desperate, path-breaking journey to the east coast is nar-
rated in his *Narrative of an Expedition to the East Coast of Greenland*.
As Holm's account of his equally pioneering expedition was never published in English, most of his findings are summarized in Thalbitzer's
brilliant *The Ammassalik Eskimo*.

Chapter Six: "That Cat's on the Roof Again"

The account of the Northern Journey is almost entirely from *Northern
Lights*, with useful assists from *Those Greenland Days* and Scott's bio of
Watkins. For the journey to establish the Ice Cap Station, Lindsay and
Scott are the sources, and Lindsay is the main source for the initial stint
with Riley manning the remote camp, with a few added diary entries
from Jonathon Riley's *From Pole to Pole*. For the prehistory of Kanger-
lussuaq Fjord, see the Wikipedia entry "Kangerlussuaq Fjord, East
Greenland" (https://en.wikipedia.org/wiki/Kangerlussuaq_Fjord,_
East_Greenland).

Chapter Seven: Autumn with the Inuit

Again, mostly Chapman and Lindsay. Almost the only account of the
ill-starred Southern Journey, but a rich one, appears as a separate chapter in Scott's bio. (The short chapter Scott contributed to *Northern Lights*
is a bare summary in comparison.) Peter Freuchen's account of falling
in love with and marrying Navarana, as well as his analysis of Inuit
sexual mores and practices, can be found in his immensely readable
Book of the Eskimos. The insider knowledge of what really went on with
Lemon, Watkins, and Chapman in their liaisons with Arpika, Tina, and
Gitrude emerges from Barker's bio of Chapman, and also in the sometimes gossipy latter-day books by Jeremy Scott and Simon Courtauld.

Interlude: The Cosmos of the East Coast Inuit

For the BAARE team's perceptions of Inuit life, Chapman and Lindsay are again the main sources. The summary of Holm's seminal study of the East Coast Inuit, as well as a trenchant critique of Holm, can be found in Thalbitzer's *The Ammassalik Eskimo*, one of the great ethnographic works of its era. That treatise also unfolds Thalbitzer's own evolving study of the natives, far ahead of its time in its cultural relativism.

Victor's brilliant, nuanced account of East Coast Inuit life in the 1930s, based on his year's habitation with an extended family in an isolated fjord north of Angmagssalik, appears in *Boréal et Banquise* (*Boréal*—the first half—alone translated into English as *My Eskimo Life*). The unbearably grim account of Inuit famine and starvation in 1882–83 lies solely in the pages of Victor's *The Great Hunger*.

Chapter Eight: "All, All Alone": Courtauld on the Ice Cap

The chief source for the first relief party's ordeal is *Northern Lights*, with valuable additions from Scott's ever-comprehensive bio. Chapman is a good source for Bingham and D'Aeth's spell at the Ice Cap Station, but the quotes from Bingham's diary come from Scott's *Portrait of an Ice Cap*. For the second relief expedition, Chapman again. But the "secret" motivation for Courtauld's volunteering to man the station solo—based on his dislike of Wager—is revealed in Wollaston's bio of Courtauld. The few quotes from Wager's diary should also be credited to Wollaston.

The coverage of the first months of Courtauld's solitary vigil derives from Wollaston's bio and from the diary quotes in Scott's *Portrait*.

Chapter Nine: Winter with the Inuit

The details of camp life in and around the hut and the interactions with the local Inuit are detailed in Chapman and Lindsay, and in Scott's bio. The extended discussion of Gino's leadership style is cogently laid out in a chapter in the bio titled simply "Leadership."

As in chapter eight, Wollaston's bio and Scott's *Portrait* illuminate the slow spiral downward of Courtauld's spirit as the months alone stretched on, vividly revealed in his diary entries.

Chapter Ten: Courtauld in Purgatory

The account of the dreadful ordeal and failure of the third relief party unfolds day by day in Lindsay's *Those Greenland Days* and Scott's bio, with a few quotations from Riley's diary taken from Jonathon Riley's *From Pole to Pole.* For Courtauld in his last weeks alone, including the revelation that he is trapped inside the domed tent, once again Wollaston's bio and Scott's *Portrait*. The retrospective comment, in which Courtauld insists that he never doubted the arrival of a relief party, appears in the chapter he contributed to *Northern Lights,* in which he consistently downplays the drama of his entrapment.

Chapman is the main source for the heroic effort of the fourth relief party led by Gino.

Chapter Eleven: Asking for Trouble: Gino's Finale

The "art of kayaking" is spelled out lovingly in *Northern Lights,* as is the narrative of the two failures on Mont Forel, as well as the rationale for the last three mini-expeditions. Scott wrote the chapter on the Ivigtut Traverse, while Chapman narrates in the third person the far dicier Holsteinborg Traverse. Courtauld's gradual change of heart after the ice cap rescue, and the quotes from his letters and diary, appear in Wollaston's bio. Percy Lemon wrote the chapter of *Northern Lights* about the Open-Boat Journey, but he minimizes completely the conflict among its three personae. For the full story, one must turn to the account Courtauld contributed to Scott's bio of Gino, and to Wollaston's bio.

Epilogue: To Slip Betimes Away

The dispiriting year between the BAARE and the vastly truncated return to Greenland in 1932, with the collapse of Gino's grand plans for

Antarctica, is sympathetically rendered in Scott's bio—which also records his meeting Margaret Graham and proposing marriage to her. The reaction of the RGS luminaries to Gino's presentation of the BAARE appears in Watkins, D'Aeth, Riley, et al., "The British Arctic Air-Route Expedition (continued)," in *The Geographical Journal*. Courtauld's extended honeymoon to the Sudan with Mollie is told by Wollaston in his bio. And Chapman's indecisive year at home comes to life in Barker's bio. Lemon's suicide is covered in Scott's bio, and more fully in Simon Courtauld's *The Watkins Boys*.

For the early days of the so-called Pan Am Expedition, including Chapman's moving account of Gino's disappearance and death, turn to Chapman's *Watkins' Last Expedition*. Scott's declining to join the expedition is somewhat hostilely explicated in his son Jeremy's *Dancing on Ice*. Scott covers the posthumous panegyrics to Gino in the bio. The sad aftermath for Margy Graham, including learning that some of Gino's friends doubted he was serious about marriage, comes mainly from *The Watkins Boys*.

For the subsequent exploratory deeds of Gino's men, consult the following:

(a)   Wager's brilliant performance on Everest in 1933 is covered in all the many histories of the world's highest mountain, as well as Hugh Rutledge's official account.

(b)   Lindsay writes vividly about his west-to-east traverse of the ice cap in *Three Got Through*.

(c)   Courtauld's successful ascent of Gunnbjørn Fjeld, the highest peak in the Arctic, is thoroughly reported in Wollaston's bio and in Courtauld's characteristically modest paper for *The Geographical Journal*, "A Journey in Rasmussen Land."

(d) Rymill's monumental three-year Graham Land expedition to Antarctica finds a thorough recounting in his *Southern Lights*.

(e) Chapman's astonishing later career as a mountaineer is well evoked in his *Living Dangerously*. His even more astonishing guerrilla campaign behind Japanese lines in World War II emerges in his classic *The Jungle Is Neutral*. Both are supplemented by Barker's bio.

The later years and often unhappy endings of the BAARE men find voice in Barker's and Wollaston's bios, in *The Watkins Boys*, in Jonathon Riley's *From Pole to Pole*, and in the case of Scott, in *Dancing on Ice*.

BIBLIOGRAPHY

Alley, Richard B. *The Two-Mile Ice Machine: Ice Cores, Abrupt Climate Change, and Our Future.* Princeton, NJ: Princeton University Press, 2000.

Amdrup, G. C. "The Danish East Greenland Expedition in 1900," *The Geographical Journal,* vol. 16, no. 6 (December 1900).

Appolonio, Spencer. *Lands that Hold One Spellbound: A Story of East Greenland.* Calgary, AB: University of Calgary Press, 2008.

Barker, Ralph. *One Man's Jungle: A Biography of F. Spencer Chapman DSO.* London: Chatto & Windus, 1975.

Bown, Stephen R. *White Eskimo: Knud Rasmussen's Fearless Journey into the Heart of the Arctic.* Boston: Da Capo Press, 2013.

Chapman, F. Spencer. *The Jungle Is Neutral.* Singapore: Times Books International, 1997 [originally 1948].

———. *Living Dangerously.* London: Chatto & Windus, 1953.

———. *Northern Lights: The Official Account of the British Arctic Air- Route Expedition, 1930–1931.* London: Chatto & Windus, 1932.

———. *Watkins' Last Expedition.* London: Chatto & Windus, 1934.

Close, Charles, and J. M. Scott, et al. "River Exploration in Labrador by Canoe and Dog Sledge: Discussion," *The Geographical Journal,* vol. 75, no. 2 (February 1930).

Conway, Martin. *No Man's Land: A History of Spitsbergen*. Cambridge: Cambridge University Press, 1906.

Courtauld, Augustine. "A Journey in Rasmussen Land," *The Geographical Journal*, vol. 88, no. 3 (September 1936).

Courtauld, Simon. *The Watkins Boys*. Norwich, UK: Michael Russell Publishing Ltd., 2010.

Freuchen, Peter. *Arctic Adventure: My Life in the Frozen North*. New York: Farrar & Rinehart, 1935.

———. *Book of the Eskimos*. Cleveland: The World Publishing Company, 1961.

———. *I Sailed with Rasmussen: Freuchen's Own Story of the Great Explorer*. New York: Julian Messner, Inc., 1958.

Fristrup, Børge. *The Greenland Ice Cap*. Copenhagen: Rhodos, 1966.

Gad, Finn. *The History of Greenland: I: Earliest Times to 1700*. London: C. Hurst, 1970.

———. *The History of Greenland: II: 1700 to 1782*. London: C. Hurst, 1973.

Georgi, Johannes. *Mid-Ice: The Story of the Wegener Expedition to Greenland*. London: Kegan Paul, Trench, Trubner & Co., 1934.

Gessain, Robert. *Ammassalik: Ou la Civilisation Obligatoire*. Paris: Flammarion, 1969.

Graah, Wilhelm August. *Narrative of an Expedition to the East Coast of Greenland*. London: J. W. Parker, 1837.

Greene, Mott T. *Alfred Wegener: Science, Exploration, and the Theory of Continental Drift*. Baltimore: Johns Hopkins University Press, 2015.

Holm, Gustav, and V. Garde. *Beretning om Konebaads-Expeditionen til Grønlands Østkyst 1883–85. Copenhagen: Meddelelser om Grønland, 1889.*

"Kangerlussuaq Fjord, East Greenland," https://en.wikipedia.org/wiki/Kangerlussuaq_Fjord,_East_Greenland.

Kintisch, Eli. "Why Did Greenland's Vikings Disappear?" https://www.science mag.org/news/2016/11/why-did-greenland-s-vikings-disappear.

Knuth, Egil. *Islandis: Au Groenland avec Paul-Émile Victor.* Paris: Éditions Paulsen, 2013.

Lindsay, Martin. Sledge: *The British Trans-Greenland Expedition, 1934.* London: Cassell and Co., 1935.

———. *Those Greenland Days.* Edinburgh: William Blackwood & Sons Ltd., 1932.

———. *Three Got Through: Memoirs of an Arctic Explorer.* London: The Falcon Press Ltd., 1946.

MacDonald, Martha, ed. *Very Rough Country: Proceedings of the Labrador Explorations Symposium.* Happy Valley-Goose Bay, Canada: Labrador Institute, 2010.

Malaurie, Jean. *The Last Kings of Thule.* New York: E. P. Dutton, Inc., 1982.

McEwen, Alec. "Labrador Boundary Dispute," https://www.thecanadiancy clopedia.ca/en/article/labrador-boundary-dispute.

Merrick, Elliott. *True North: A Journey into Unexplored Wilderness.* Berkeley, CA: North Atlantic Books, 2010 [originally 1933].

Mikkelsen, Ejnar. *Two against the Ice.* South Royalton, VT: Steerforth Press, 2003.

Millman, Lawrence. *A Kayak Full of Ghosts: Eskimo Folk Tales.* Northampton, MA: Imprint Books, 2010.

———. *Last Places: A Journey to the North.* Boston: Houghton Mifflin Company, 1990.

———. *Lost in the Arctic: Explorations on the Edge.* New York: Thunder's Mouth Press, 2002.

Nansen, Fridtjof. *The First Crossing of Greenland.* London: Longmans, Green, and Co., 1890.

"The Prehistory of Greenland," https://natmus.dk/organisation/forskning -samling-og-bevaring/nyere-tid-og-verdens-kulturer/ etnografisk-samling/arktisk-forskning/prehistory-of-greenland/.

Priestley, Raymond. *Antarctic Adventure: Scott's Northern Party.* London: T. F. Unwin, 1914.

Rasmussen, Knud. "Report of the Second Thule Expedition for the Exploration of Greenland from Melville Bay to De Long Fjord, 1916–1918," Copenhagen: *Meddelelser om Grønland*, vol. 65 (1927).

Riley, Jonathon P. *From Pole to Pole: The Life of Quintin Riley, 1905–1980.* Bluntisham, UK: Bluntisham Books, 1989.

Rymill, John. *Southern Lights: The Official Account of the British Graham Land Expedition 1934–1937.* London: The Travel Book Club, 1939.

Scott, J. M. *Gino Watkins.* London: Hodder and Stoughton Ltd., 1935.

———. *The Land That God Gave Cain: An Account of H. G. Watkins' Expedition to Labrador, 1928–1929.* London: Chatto & Windus, 1933.

———. *Portrait of an Ice Cap: With Human Figures.* London: Chatto & Windus, 1953.

———. *The Private Life of Polar Exploration.* Edinburgh: William Blackwood, 1982.

———. *Unknown River.* London: Hodder and Stoughton, 1939.

Scott, Jeremy. *Dancing on Ice: A Stirring Tale of Adventure, Risk and Folly.* London: Old Street Publishing, 2008.

Sonne, Birgitte. *World-Views of the Greenlanders: An Inuit Arctic Perspective.* Fairbanks: University of Alaska Press, 2017.

Thalbitzer, William, ed. *The Ammassalik Eskimo: Contributions to the Ethnology of the East Greenland Natives.* Copenhagen: Bianco Luno, 1923.

———. *The Heathen Priests of East Greenland (Angakut).* Vienna: A. Hartleben, 1909.

Victor, Paul-Émile. *Boréal et Banquise.* Paris: Bernard Grasset, 1939.

———. *Chants d'Ammassalik.* Copenhagen: Meddelelser om Grønland, Man and Society 16, 1991.

———. *The Great Hunger [La Grande Faim]*. London: Hutchinson, 1955.

———. *Les Survivants du Groenland*. Paris: Éditions Robert Laffont, 1977.

———. *My Eskimo Life*. New York: Simon and Schuster, 1939.

Victor, Paul-Émile, and Joelle Robert-Lamblin. *La Civilisation du Phoque: Jeux, Gestes et Techniques des Eskimo D'Ammassalik*. Saint-Quentin-en-Yvelines, France: Armand Colin and Raymond Chabaud, 1989.

Watkins, H. G. "The British Arctic Air-Route Expedition," *The Geographical Journal*, vol. 79, no. 5 (May 1932).

Watkins, H. G., and N. H. D'Aeth, Quintin Riley, et al. "The British Arctic Air-Route Expedition (continued)," *The Geographical Journal*, vol. 79, no. 6 (June 1932).

Watkins, H. G., and H. T. Morshead, R. Woolley, et al. "The Cambridge Expedition to Edge Island," *The Geographical Journal*, vol. 72, no. 2 (August 1928).

Watkins, H. G., and J. M. Scott. "River Exploration in Labrador by Canoe and Dog Sledge," *The Geographical Journal*, vol. 75, no. 2 (February 1930).

Waugh, Evelyn. *A Little Learning: The First Volume of an Autobiography*. Boston: Little, Brown and Company, 1964.

Wegener, Else, ed. *Greenland Journey*. London: Blackie and Sons Limited, 1939.

Whipplesnaith [Noel H. Symington]. *The Night Climbers of Cambridge*. Cambridge, UK: The Oleander Press, 2007 [originally 1937].

Wollaston, Nicholas. *The Man on the Ice Cap: The Life of August Courtauld*. London: Constable, 1980.

INDEX

Watkins, Henry George "Gino," (*continued*)
  temper of, 206
  third relief mission and, 205
  tone set by, 200
  wireless message on Courtauld's rescue, 234–35, 243
Watkins, Henry George (senior), 15–17, 19–20, 21, 22, 23, 39–40, 47–48, 76, 275, 277, 301
Watkins, Pamela, 47–48, 74, 275, 276, 286, 288, 301, 314
Watkins, Tony, 47–48, 74, 288, 301
Watkins Mountains, 309–11
Watkins Mountains expedition, 309–11
Waugh, Evelyn, 18
weather observations, 287
Weddell ice sheet, 280
Weddell Sea, 281

Wegener, Alfred, 242n
Welsh Harp, 279
Western Settlement, 96, 98, 99, 102
West Greenlanders, 99, 100, 119
whaleboats, 259, 261
whale hunting, 31, 32, 97
whale meat, 87, 89
Whipplesnaith, 24
Wiessner, Fritz, 313
Wild, Frank, 306
Wilkins, Hubert, 312
Winnipeg, 4
wireless, 172, 174, 185–86, 213, 262, 274
Wollaston, Nicholas, 184
wolves, 107
Wordie, James, 28, 29, 31, 87–88, 161, 184, 279, 280, 281, 306
World War I, 107
World War II, 313–14, 316
Worsley, Frank, 28